AN ILLUSTRATED A-Z HISTORY OF
TRACTORS

LORENZ BOOKS

AN ILLUSTRATED A-Z HISTORY OF
TRACTORS

Mirco De Cet

This edition is published by Lorenz Books
an imprint of Anness Publishing Ltd
108 Great Russell Street
London WC1B 3NA
info@anness.com

www.lorenzbooks.com
www.annesspublishing.com

Anness Publishing has a new picture agency outlet
for images for publishing, promotions or advertising.
Please visit our website www.practicalpictures.com
for more information.

© Anness Publishing Ltd 2014

A CIP catalogue record for this book
is available from the British Library.

Publisher: Joanna Lorenz
Senior Editor: Felicity Forster
Produced by Editorial Developments,
    Edgmond, Shropshire, England
Production Controller: Pirong Wang

PUBLISHER'S NOTE
Although the information in this book is believed
to be accurate and true at the time of going to press,
neither the authors nor the publisher can accept
any legal responsibility or liability for any errors
or omissions that may have been made.

Contents

Introduction

At one time or another, we have all encountered a tractor, often down some tiny lane in the middle of nowhere!

Meeting one of these huge, modern monsters, with their large wheels towering over you, the driver tucked up in a high air-conditioned cab, probably listening to music on a sophisticated hi-fi system, can be a pretty daunting experience. No longer do tractors travel at low speeds, and no longer can you drive past one with ease. The modern tractor can be as comfortable and as easy to drive as a modern car. Long gone are the days of hard metal seats, heavy steering and machines that are open to the elements.

Men and women toiled long and hard in the fields, to grow crops that would feed their families. With the advent of implements that could be pulled by animals, things improved. Then came mechanization – first it was steam-powered engines, working alongside horses and bulls, and then gradually the animals were replaced with more sophisticated machinery.

In 1892 the development of the tractor took a huge leap into the future. John Froelich built what is reputed to be the first gasoline-powered farm vehicle – the word 'tractor' was as yet not invented – and, along with his Waterloo Gasoline Traction Engine Company, produced a further series of machines. The Waterloo Company was bought out by John Deere, plunging them headlong into the delicate tractor manufacturing market. Today, John Deere is one among an elite of tractor manufacturers who dominate the farming community worldwide.

Henry Ford also made his name in the tractor business, and, as with his cars, produced tractors at highly competitive prices – by 1925 Ford owned more than half of the tractor market of the day, selling some 100,000 Fordsons per year – mass production of the tractor had *truly arrived. As the 1930s dawned, so tractors became sleeker and cleaner looking. It was now, too, that the use of rubber tyres was introduced, which must have been a godsend for the driver, who spent hours – possibly days – on his tractor. The smooth ride, better fuel economy and better grip only went to enhance production and comfort.*

As the tractor and all its implements became more refined and sophisticated, so larger and larger areas of crops were being harvested in an ever shorter space

of time. Today's tractors are more like moving electronic workstations, with their hi-tech cabs filled with the latest technology, calculating every move being made, and every move that needs to be made. Benjamin Franklin once said, "Time is money", and today that phrase is positively suited to the work of the tractor.

AGCO

Even though the company had been bought out, this tractor clearly shows its Allis-Chalmers heritage through its orange bodywork. This AGCO 6690 was made between 1991 and 1997.

Although the AGCO (Allis-Gleaner Corporation) has roots that can be traced all the way back to the mid-19th century, when it comes to building tractors, they are actually a new company.

In 1985 Allis-Chalmers was purchased by Deutz of Germany, who closed down the factory and commenced exporting tractors to the USA under the Deutz-Allis name. The idea was to build their own tractors in the USA, but finally they contracted the White company to build for them, using Deutz air-cooled diesel engines. This project was short-lived, and within a year the US Deutz-Allis management bought itself out, and the AGCO (Allis-Gleaner Corporation) was born. The tractors were now named AGCO-Allis, and painted in the familiar Allis orange.

The following years saw a huge frenzy of purchases: in 1991 AGCO purchased the Hesston Corporation, a leading American brand of hay tools, along with a 50 per cent stake in a manufacturing joint venture with Case International, known as Hay and Forage Industries (HFI). It also purchased the White tractor business – giving the company its first tractor-making facility.

In 1992, the company presented a range of 15 tractors, of which 12 were painted orange,

In 1984, Allis-Chalmers found themselves in difficulty. The company was purchased by Deutz, who rebranded the Allis tractors as Deutz-Allis.

Shown here with rear grass-cutting attachment is the AGCO-Allis two-wheel drive 5670. It was manufactured by SLH (SAME/Lamborghini/ Hurlimann), and used their four-cylinder diesel engine.

The Model 8630 AGCO (above) was also manufactured by SLH on their behalf. This one used the six-cylinder, turbo-diesel engine, and had a capacity of 5506cc (336 cubic inches).

The AGCO 9650 used a Deutz six-cylinder, turbo-diesel unit, and had 18 forward and nine reverse gears. This model was made in 1995.

having been made for AGCO by SLH in Italy. These machines had AGCO-Allis badges, while the remainder were painted silver and bore the White badge.

In 1994, under the symbol AG, AGCO was listed on the New York stock exchange. In 1993 they purchased the White-New Idea business of planters, hay tools and spreaders, along with the Coldwater, Ohio manufacturing facility. It also purchased the US distribution rights to Massey Ferguson products, expanding the AGCO US dealer network by over 1,000. The remainder of the Massey Ferguson worldwide holdings was snapped up in 1994, ensuring AGCO global status.

FACT BOX

AGCO products

- **Massey Ferguson** has more than 157 years of innovation and experience, and is the most widely sold tractor brand in the world.
- **Fendt** is a market leader in Europe, with a distinguished reputation for superior technology and engineering.
- **Valtra** is based in Finland and is well established in Brazil's rapidly growing market.
- **Challenger** uses a revolutionary track system.
- **Hesston** has been the hay and forage innovator since 1955.
- **Gleaner** introduced the first self-propelled combine in 1923.

For nearly 40 years, Spra-Coupe has been producing high-performance row-crop sprayers. Powerful diesel engines provide plenty of power for these field-friendly, low-impact sprayers.

The Challenger MT models are specifically designed as tracked tractors, with transmissions and engines tailor-made to handle the high draft loads that tracks produce.

AGCO purchased the Ag-Chem Equipment Company, Inc. in 2001. These are machines built on rugged frames and engineered to withstand rough field work and tricky terrain.

McConnell tractors was also bought, leading to the development of the AGCO-Star articulated tractor line and Black Machine. In 1995 AGCO purchased the AgEquipment Group, makers of Glencoe tyre and farmhand agricultural implements and tillage equipment.

Finally in 1996 the Deutz diesel engine era came to an end, but there were further acquisitions of the Iochpe-Mexion agricultural equipment

The 235hp Challenger MT585D, introduced in 2011, uses the Tier 4 interim emissions-compliant AGCO engine with clean air technology.

The 2013 tracked MT800D offers a cab packed with the latest technology for optimizing performance for both driver and machine.

company in Brazil, Deutz in Argentina, and the Western Combine Corporation and Portage Manufacturing Inc. in Canada. The following year AGCO acquired Fendt GmbH, a leading German tractor business, while in 1998 a joint venture was created with Deutz AG to produce engines. Spra-Coupe and Willmar, leading companies in the agricultural sprayer market, were also purchased.

Two US plants closed in 2000, and the Massey Ferguson plant in Coventry, UK, ceased production in 2003. For 2001 the Ag-Chem Equipment Company was purchased, and a year later they bought the design, assembly and marketing of Caterpillar's Challenger tractors. Valtra tractors from Finland was purchased in 2004 and became one of four core brands with Challenger, Massey Ferguson and Fendt. In 2011 they added grain storage company GSI to their portfolio of 20 brands. With manufacturing in 25 global locations and sales of nearly US$10 billion in 2012, AGCO is a world leader in the manufacture and distribution of agricultural equipment.

Allis-Chalmers

By 1979, Allis-Chalmers had grown into a US$2 billion corporation and one of the most important machinery and tractor manufacturers in the United States – but its origins were humble.

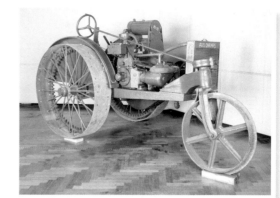

The Allis-Chalmers 10-18 was basically an engine mounted on a frame, which was cooled via a radiator. The tricycle-style layout allowed for a tighter turning radius.

The Allis-Chalmers Model U was initially known as the "United" and was produced for The United Tractor & Implement Company in 1928.

After the formation of the Allis-Chalmers Company in 1900, it continued to concentrate on heavy engineering, but as the US economy hit hard times so did the company, and in 1913 receivers were brought in to carry out a re-organization. One of these was General Otto Falk, who became president of Allis-Chalmers, and its saviour. With a personal interest in farming, he decided to diversify into tractors, and with World War I making heavy demands on agriculture, this sector was experiencing a period of sustained growth.

A couple of machines were tested and put up for sale prior to the Model 10-18 making its appearance, of which small-scale production was started in 1914. A horizontally opposed, twin-cylinder engine could be started on petrol and then switched to paraffin. The 10-18 continued for some years and was eventually replaced by the 12-20 in 1921. 1918 saw the production of the 6-12, a motor plough designed to be hooked up to existing horse-drawn ploughs or binders.

Allis-Chalmers set up a new tractor division in 1918 and the company now produced the 15-30, a conventional machine with four wheels and a four-cylinder engine. Although powerful, it was still a heavy tractor and had a large price tag. In an attempt to compete in the mid-range tractor market, Allis-Chalmers produced the smaller 12-20 machine, later redesignated the 15-25.

With low sales figures and machines that were too expensive and too heavy, Allis-Chalmers found itself in trouble. It was now that Otto Falk promoted Harry Merritt to head of the tractor

This is one of several Allis-Chalmers badges. This was rarely seen on the radiator, where just the company name was written.

Allis-Chalmers

FACT BOX

Formation of Allis-Chalmers

- **1857** Reliance Works collapses during a financial panic.
- **1860** Edward P. Allis purchases Milwaukee's Reliance Works.
- **1861** Edward P. Allis & Company founded by E. P. Allis of New York, following his acquisition of Reliance Works.
- **1869** The company expands into steam power; the first Allis steam engine makes its debut.
- **1877** Edwin Reynolds, a steam expert, joins the firm.

Company moves into the West Allis works, where tractors are made until 1985.
- **1889** Edward Allis dies. The company continues to manufacture under the supervision of Edwin Reynolds.
- **1900** A chance meeting between Reynolds and William Chalmers, of Fraser and Chalmers Company, leads to the formation of Allis-Chalmers in 1901.

In 1928 Allis-Chalmers bought out the Monarch Tractor Company of Springfield, Illinois, who made crawlers. A direct result was this Model K.

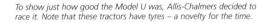

To show just how good the Model U was, Allis-Chalmers decided to race it. Note that these tractors have tyres – a novelty for the time.

division, who instigated huge price reductions. The 20-35, for example, was offered at half its original selling price and so the machines started selling well. When redesignated Model E, it sold over 19,000 units in a 20-year period.

When Henry Ford decided to pull out of tractor manufacturing in the USA, a group of Fordson dealers decided to form an association. Allis-Chalmers was invited to join them, and so the United Tractor and Equipment Company of Chicago came into being. Besides other farm machinery, it wanted to produce a medium-sized tractor, which Allis-Chalmers would make for them. When introduced in 1929, it was marketed as the United and used a Continental engine, but before long the consortium ran into difficulties. Allis-Chalmers took on the marketing as well as the production of the tractor, which now became the Model U, later fitted with an Allis-Chalmers paraffin engine. Allis-Chalmers continued to expand, buying out the Advance-Rumely concern in 1931.

Much work was being done at this time to improve tractor wheels, which were still made of steel with lugs or cleats. All types of experiments were carried out, but a breakthrough came in 1932, when aircraft tyres were tested. Later in the

The Model A tractor was produced between 1936 and 1943, and could be bought with either steel wheels or rubber tyres. It ran on petrol.

year the first tractor with rubber tyres went on sale – although there was some reluctance to accept the change. So Allis-Chalmers decided to race the tractors as a publicity stunt. The steel-wheel Model U had a top speed of about 6kph (3.7mph), but when an extra gear was added to the rubber tyre model, it could reach 16kph (10mph). The Model U was also modified for racing and became the fastest tractor in the world. It caused a sensation during 1933 after demonstrations at state fairs, and by 1937 almost 50 per cent of tractors sold were fitted with rubber tyres.

The Model WC of 1933 was to become one of the company's best-selling machines. With its water-cooling, four-stroke engine and four-speed transmission, it could reach just over 14kph (8.6mph) on rubber tyres. So popular was the machine that it went on selling through to the late 1940s, final sales reaching some 178,000 units.

An Allis-Chalmers advertisement showing how a family could all be involved with the everyday running of their farm, and the versatility of the tractor.

Built between 1941 and 1949, this is an example of a row-crop Model C. Its four-cylinder engine had a displacement of 2048cc (125 cubic inches).

Allis-Chalmers

Close-up view of the Allis-Chalmers Model B petrol engine. Initially a Waukesha petrol engine was supplied for the first 96 tractors of 1937.

The Model G Allis-Chalmers had its Continental engine positioned at the rear. It was aimed at the smaller farmer and market gardeners.

The 1930s saw the introduction of the Model A – effectively a replacement for the 18-30 machine – and then the Model B of 1937 – an ultra-lightweight machine using unit construction and four-cylinder engine. Designed by industrial designer Brooks Stevens, it came with a low price tag, which made it an affordable tractor for many of the smaller-capacity farms. This was followed by an uprated version, the Model C, which also sold well. By now, Allis-Chalmers had established itself as a major player in the tractor market.

Allis-Chalmers opened a factory in the UK, and from 1947 the Model B was made in a plant at Totton, near Southampton. Initially it was assembled from components shipped over from the USA, but later used locally sourced parts, finally becoming completely UK-built at Essendine in Leicestershire. There was a Perkins P3 diesel option with four-speed transmission and hydraulics, which made a distinct improvement. When the Model B was no longer competitive and production ceased, it was replaced by the D270, and later by the D272.

The last Allis-Chalmers model to appear in the UK was the ED40 in 1960, which used a Standard-Ricardo 2.3 litre engine, and featured an eight-speed gearbox, live hydraulics, and the option of live PTO. Depth-O-Matic hydraulics were introduced in 1963, as was an engine power increase from 37 to 41hp.

A view along the Model B's font end, showing its four-cylinder petrol engine. It was made between 1938 and 1957, and was attractive for its size and price.

Allis-Chalmers

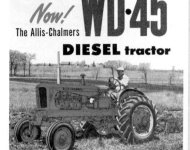

Now! The Allis-Chalmers WD·45 DIESEL tractor
with AC FULL DIESEL ENGINE

A period advertisement for the Allis-Chalmers WD45, pointing out that it now had a full diesel engine.

Produced from 1953, the WD45 had the same general look as the previous WD tractors, but was much more powerful due to its new four-cylinder 'Power Crater' petrol engine.

Just before World War II, the WF model was presented, but unfortunately production stopped due to material shortages caused by the ongoing conflict. Production restarted at the end of the war, and by 1948 most of the pre-war features, including electric lights and starter, were back on the market. The Model G – reminiscent of the early Allis-Chalmers motor plough – made its appearance using a small four-cylinder Continental engine.

In 1948 the WD45 was introduced, which had the same general look as the WD tractor but was more powerful, having the new four-cylinder Power Crater petrol engine. In 1954 the WD45 was offered with the Snap-Coupler system, and was also the first Allis-Chalmers to have power steering. The end of production for the Models B and G saw the introduction of the D series. The first to appear were the D14

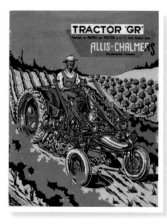

TRACTOR 'GR'
ALLIS-CHALMERS

A French advertisement of the period, showing a happy farmer on his Allis-Chalmers GR, which was aimed at smaller holdings.

Allis-Chalmers

The Model D10 was manufactured between 1959 and 1967. It used a four-cylinder petrol engine, initially of 2261cc (138 cubic inches), subsequently uprated to 2441cc (149 cubic inches).

This is a fine example of a One-Ninety, which used the larger six-cylinder petrol engine. This is a turbocharged version, denoted by the XT badge.

and D17, which became replacements for the WD model, and were three-plough and four/five plough models respectively. The diesel or LPG-powered D17 was one of the company's most popular tractors and featured a power-director hand clutch, power-adjustable rear wheels and traction-booster system. For 1959 they were joined by the D10 and D12 – the smallest of the new models. Both used a water-cooled, four-cylinder engine and were never fitted with diesel engines.

The late 1950s was also a period of great experimentation, and Allis-Chalmers dabbled with the idea of a fuel-cell tractor, which used a Model D12 chassis fitted with 1,008 individual cells fuelled by a mixture of gases – mostly propane – which in turn created a current flow that was channelled through to an Allis-Chalmers 20hp direct-current electric motor.

With Allis-Chalmers models once again falling behind the competition, attempts were made to increase the power of current machines. The D18 was upgraded to the D19 with a 17hp petrol engine, and there was also a turbocharged diesel version – a first for the farming world. These machines were followed by the company's first 100hp model, the D21, which used a large-capacity, direct-injection, six-cylinder diesel engine. This tractor was so powerful that not only did it

Allis-Chalmers

Weighing in at 9,900kg (21,826lb), this is the six-cylinder, turbocharged 4W-305 model tractor, produced from 1982 to 1985. It was ideal for the heavyweight jobs around the farm.

need a completely new eight-speed transmission, but also new implements were designed to suit, such as a seven-bottom plough.

A new 100 series arrived for the 1960s, continuing the rather squared-off styling of the D21, its replacement being the One-Ninety. One-Ninety was found lacking, so Allis-Chalmers decided to add a turbocharger and changed the designation to One-Ninety XT. The D17 was replaced by the One-Seventy, while the One-Eighty had the option of a cab. They were similar machines and benefitted from a new family of six- and four-cylinder engines, with the One-Seventy also having a Perkins diesel option. For machines with smaller engines, the company turned to Fiat of Italy, having already established links with the

The 6140 model, suited to the smaller workload, used a Toyosha three-cylinder diesel engine. It was one of the last tractors made prior to the closure in 1985.

Produced between 1960 and 1967, the D15 had a choice of petrol, LP gas and diesel engines, and was classed as a mid-range tractor.

Seen here is the Model 8050, a six-cylinder, turbo-diesel machine, with a displacement of 6980cc (426 cubic inches). It had 12 forward and two reverse gears, and independent PTO.

The slightly smaller 8030 used the same engine as its bigger stable mate, the 8050. The 7000 series tractors had been replaced by the 8000 in 1982.

company. The 5040 had a Fiat three-cylinder diesel engine and bigger versions, 5045 and 5050 – shortly replaced by the 6000 series – all used Fiat engines too.

The early 1970s' big 7000 Power Squadron machines used Allis-Chalmers 7 litre diesel engines, producing 130hp when turbocharged, and 156hp when turbo-intercooled. The lower-powered 7000 series followed, including the 7080 four-wheel drive model, which in turn was replaced by the larger 8550.

By the early 1980s the company was finding the economic situation difficult, following a huge downturn of sales both in the USA and Europe. The last tractors bearing the Allis-Chalmers name were produced in the USA in December 1985, the farm equipment division being sold that year to K. H. Deutz AG of Germany. The Allis name has survived into the 21st century under the AGCO flag, but Allis-Chalmers is no more.

The 5000 series tractors began using three-cylinder Fiat engines, and when the 6000 range took over, they too used these engines. This is the 6080.

Belarus

The Production Association of Minsk Tractor Works (P/A MTW), the main tractor plant of the former Soviet region of Belarus (now a republic), was founded on May 29, 1946. It was a true association on the communist model.

The 1500 model was a four-wheel drive machine, and was equipped with a powerful turbo-diesel engine.

Shown here is the Belarus S 920.3, a versatile and economical 100hp, four-wheel drive tractor with a fourteen forward and four reverse, synchromesh gearbox.

БЕЛАРУС®
ПО "МИНСКИЙ ТРАКТОРНЫЙ ЗАВОД"

It consisted of a Special Tools and Production Equipment Plant, the Vitebsk Tractor Spare Parts Plant, the Bobruisk Tractor Parts and Units Plant, and the Specialized Engineering Bureau for Versatile Row-Crop Tractors. It became one of the world's largest manufacturers of agricultural equipment, employing nearly 20,000 workers, producing over three million tractors, and exporting some 500,000 of these to more than 100 countries.

Production started with the KD-35 tracked model, followed in 1953 by the MTZ-2 tractors, which used pneumatic tyres. Then came the odd-looking KT-12 tracked model, and by 1958 MTW had made its 100,000th machine. Production of the MTZ-50 versatile wheeled tractor was launched in 1961, and the MTZ-52 was introduced shortly afterwards. In 1974 MTW started bulk production of the much more powerful MTZ-80 model, which was to become the biggest selling tractor in the world.

MTW started its development of mini tractors in 1978, creating and producing motoblocks (walk-behind

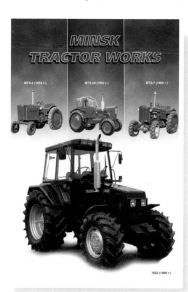

An advertisement for the Belarus tractor shows not only the new model, but the older ones too. Belarus has been making tractors for many years.

Belarus

The Belarus 952.3 is one of the latest versatile and economical 100hp, four-wheel-drive tractors that the company produces. The engine is a four-stroke, direct-injection, water-cooled, intercooled turbo-diesel unit.

A brochure depicting the Belarus 1221.3, currently the most powerful model of the range in the UK. This 140hp tractor offers great performance and economy.

This older Model 5380 stands high but gives its age away with the rather squared-off styling. All the same, it was still a powerful machine, as its advertising notes.

ПО "МИНСКИЙ ТРАКТОРНЫЙ ЗАВОД"

BELARUS®

P/A MINSK TRACTOR WORKS

type tractors), the MTZ-05, MTZ-06, MTZ-08BS and MTZ-12, along with the MTZ-082, a four-wheeled mini-tractor that used a 12hp engine. The association continued to improve its products, and 1984 saw the introduction of the MTZ-102, 100hp machine.

By March 1984, the 2,000,000th tractor left its main assembly line, while the first example of the MTZ-142, 150hp tractor was made the following year. Several versions of the MTZ-220 tractor, using a 220hp unit, were designed and mass-produced.

The 1990s were almost as busy. In 1994 MTW started production of the MTZ-1221, which used a 130hp unit, and by 1995 the 3,000,000th tractor had rolled off the line. In 1999, MTW presented a new development, the MTZ-2522, 250hp versatile tractor. Their 1802 tracked vehicle was presented in time for the anniversary of the Works in 2000.

Belarus continue to expand and innovate, with their 300hp 3023 winning awards for its electro-mechanical drivetrain system in 2009. Today they distribute to 125 countries and offer 100 tractor models, and more than 200 assembly options.

20

Case IH

Jerome Increase Case founded the J. I. Case Company in 1842, and is credited with building the first steam engine used for agriculture. Whilst president, the company manufactured more threshing machines and steam engines than any other manufacturer in history.

Jerome Increase Case, a young man in 1842, headed for Rochester, Wisconsin Territory. He had heard it was the wheat centre of the Mid-West, and planned to become a thresherman there.

The Case 12-25 was introduced in 1914. It catered for the smaller farmer, who found the larger 30-60 models inappropriate. It used an opposed two-cylinder engine.

Known in manufacturing circles as the 'Threshing Machine King', Case received more popular recognition as the owner of 'Jay-Eye-See', a black gelding racehorse acknowledged as the world's all-time champion trotter-pacer.

It was the young Case who brought his father's attention to an advertisement in the Genesee Farmer about a new machine that would thresh wheat. Since biblical times, the farmer had cut wheat with scythes, then threshed it by hand with flails, and winnowed the grain from the chaff by tossing it in the air – and it was very back-breaking work.

Case was born into a pivotal period in American agriculture, and by applying technology to farming, he, along with others, raised production levels to the extent that the United States became the breadbasket of the world.

In 1842 Case took a crude 'Ground Hog' threshing machine with him from Williamstown, New York, to Rochester, Wisconsin, where he improved its design and

Most companies have a trademark, and in 1865 the J. I. Case Company adopted the Civil War eagle 'Old Abe' as their trademark, and as a symbol of excellence.

Case IH

Seen from behind, this is the 10-20 in action. One forward gear gave a top speed of 3.2kph (2mph), and final drive was by chain.

The steering wheel of the 10-20 was mounted on the right, and the driver's seat was positioned directly behind the right rear wheel, which was keyed to the axle.

established his company. By the start of the American Civil War, the number of reapers and mowers being used on farms had grown from 90,000 to 250,000 – the war itself stimulating expansion in farm mechanization.

In 1863 Case took on three partners to create the J. I. Case Company: Massena Erskine, Robert Baker and Stephen Bull; and these men soon became known as the Big Four. Two years later, the eagle trademark was adopted. It was based on 'Old Abe', the emblem used by Company C of the 8th Wisconsin Regiment during the Civil War. In 1894 the eagle was depicted on top of a globe, and was to remain the company logo for the next 75 years.

In 1869 Old No.1, the first Case steam engine, appeared, and although mounted on wheels, it was pulled by a team of horses and was only used for belt power to other attachments – it was another 15 years before the steam boom would truly take off.

This odd-looking Case tractor is the 10-20 model, and was produced between 1915 and 1918. It used a four-cylinder, transverse-mounted engine.

Seen here is an early depiction of a farming scene, in which a Case steam engine is being used to power a threshing machine via a leather belt.

This is an excellent example of a Case 'crossmotor' tractor. As the description of the engine implies, it was mounted transversely in the engine bay.

A new line of tractors was introduced by Case in 1929. This one, known as the 26-40, and better known as the Model L, used a four-cylinder Case engine.

The first Case traction engine came in 1876, and in 1880 the J. I. Case and Company partnership was dissolved. The J. I. Case Threshing Machine Company was born, and by 1886 it was the world's largest manufacturer of steam engines.

On December 22, 1891, Jerome Increase Case died, and the town of Racine mourned one of the USA's industrial pioneers. Stephen Bull took over as president of the company. William Paterson was employed to produce a new internal combustion engine to power a tractor, but his design had an unusual layout – the engine had two opposing

This Model C was a smaller version of the Model L. These were introduced in 1929, and a further CH model – high clearance – was also produced later.

Case IH

The Model R was produced in 1938, but production shortages during the war period limited their manufacture. Post-war, there were orchard and industrial models too.

The Model RC Case tractor, a row-crop model, was introduced in 1935. It was restyled in 1939, along with the other tractors, and was fitted with rubber tyres.

pistons in one cylinder, with a complex linkage between them and the crank. The unreliability of the engine prompted the company to abandon it. In 1904 Case introduced its first all-steel thresher, and was producing more farm steam engines and threshing machines than its rivals.

In 1910 the Gasoline Traction Department demonstrated a prototype to the Case board, who gave the go-ahead to build. Later the following year, the Case 30-60 tractor was launched, with an unconventional horizontal twin-cylinder engine unit. A smaller 20-40 model with flat-twin engine followed, and in 1913 Case produced the 12-25.

That year the Case Tractor Works (known for a time as the Clausen Plant) was built near Racine to manufacture several sizes of four-cylinder, petrol engine tractors, including cross-mounted units – these engines being a development from the purchase of the Pierce Motor Company, which Case studied and redesigned to sit crossways in the 10-20 model tractor of 1916. The 22-40, unveiled in 1919, was the most powerful tractor the company had yet produced. Although they were expensive, with this power they could also pull Case threshers, and they more than paid for themselves. The biggest cross-motors model was the 40-72 of 1920.

By the mid-1920s, Case was struggling, and new products were needed to keep up with the

The Model D of 1939 was a four-wheeled tractor, and was designed to pull a three-tine plough. It had a belt pulley, PTO and Case's motor lift system for the implements.

competition. Leon R. Calusen was installed as president in 1924, and he started by closing the automobile section, discontinued steam engine production and requested D. P. Davies to design a new engine to take the place of the cross-motors. Two new machines arrived in 1929, the Model L and Model C, both using four-cylinder engines fitted longitudinally, with a three-speed gearbox and chain final drive.

In 1932 the little RC model was produced, which used a bought-in Waukesha engine that ran on petrol only. In 1937 The Rock Island (Illinois) Plough Company factory was purchased, and in 1939 'flambeau red' became the identifying colour for Case equipment. A new fleet of tractors was also introduced, including the Model D, which was basically a restyled C but with additions such as mechanical motor lift, disc brakes, and live PTO. The Model S, a two-plough general-purpose machine that replaced the RC, was also presented and, unlike the D, had four-speed transmission; it was available in different versions, depending on the work load. The Model V was supposedly a lightweight, but was heavier and more powerful than its competitors.

The first Case diesel was manufactured in 1953 as the 500 model. It was a six-cylinder machine and had an engine capacity of 6177cc (377 cubic inches).

The Model S, a smaller version of the D, was manufactured in 1940. Initial production was slow due to the war, but further SO (orchard) and SC (row-crop) models followed.

The VAC model, seen here, was a replacement for the VC model. It was in production up to 1955, and was equipped with lights and electric starter as standard from 1952.

Built between 1955 and 1958, the 300 series Case tractors came with a choice of Case spark ignition engine or four-cylinder Continental diesel motors.

The 400 model was a new model for 1955, and available with petrol or diesel engine. It was also equipped with an eight-speed transmission unit.

World War II saw Case supply material for the war effort – hundreds of thousands of 155mm (6.1in) shells, 40mm (1.58in) anti-aircraft gun carriages, B-26 bomber wings and after-coolers for Rolls-Royce aircraft engines. Post-war production continued with pre-war models, although by now the Model V had been replaced by the VA, fitted with the new snap-on Eagle Hitch. The 1950s saw Case introduce the six-cylinder 500 diesel model – making it one of the most powerful tractors available to the farmer.

A new president, Mark Rojtman, joined Case in 1956, when Case absorbed the American Tractor Company. Current models were already going through an updating stage, and in 1955 the 400 model was introduced with 'desert sunset' orange sheet metal, and 'flambeau red' chassis. It used a four-cylinder unit and could run on petrol, diesel or LPG. For 1956, a new 300 model was introduced which catered for the smaller farmer, and the 500 became the 600, with new six-speed transmission. A significant introduction to the models at this time was the Case-o-Matic drive system.

In 1964 the turbocharged diesel, four-wheel drive 1200 Traction King was introduced, aimed at high-production agricultural operations.

The early 1960s saw Mark Rojtman leave after a boardroom bust-up, and in 1967 the company was taken over by Tenneco Inc. of Houston, Texas, giving the company financial security and allowing

Case IH

It was 1966 when Case introduced their 1030 Comfort King model. Although not turbocharged, it was capable of pulling large loads with considerable ease.

Clearly identified here is the 600 model. This was introduced in standard track format in 1957, and used a six-cylinder Case diesel engine.

This is the Case 2670 Traction King, a monster machine that had four-wheel drive, four-wheel steer and an intercooled, turbo-diesel engine that could produce 221bhp.

Case IH

case**iii**

56 SERIES
85-105 DIN HP

WITH THE UNIQUE
SENS-O-DRAULIC
SYSTEM

An advertising brochure for the Case 56 series tractor. Note the new company logo, introduced after the merger with International Harvester.

FACT BOX

Manufacturing plants in 2006

EUROPE

Belgium, France, Germany, Italy, Poland, United Kingdom, Austria: Tractors, Components, Combines, Forage Harvesters, Large Rectangular Balers, Grape Harvesters, Wheeled Excavators, Wheel Loaders, Graders, Tractor Loader Backhoes, Mini-Excavators, Midi-Wheeled Excavators, Telehandlers, Dozers, Wheel Loaders, Crawler Excavators, Combines, Balers, Engines.

AFRICA AND ASIA

Turkey, China, India, Pakistan, Uzbekistan: Tractors, Tractor Loader Backhoes, Cotton Pickers.

LATIN AMERICA

Brazil, Mexico: Crawler Excavators, Tractor Loader Backhoes, Crawler Dozers, Wheel Loaders, Graders, Sugar Cane and Coffee Harvesters, Planters, Tractors, Combines, Tractors, Components.

NORTH AMERICA

Canada: Planting, Seeding Equipment. *United States:* Hay and Forage Equipment, Sprayers, Floaters, Cotton Pickers, Tractor Loader Backhoes, Forklifts, Crawler Excavators, Dozers, Compact Tractors, Tractors, Wheel Loaders, Soil Management (Tillage) Equipment, Combines, Components, Engines, Skid Steer Loaders.

it to continue producing its many different products. Unfortunately 'Old Abe' was replaced by a bold CASE logo, and model numbers changed – 900 becoming the 930, and so on. Much emphasis was now on sheer power, and Case introduced the 1030 Comfort King model, which produced 102hp from its water-cooled, six-cylinder engine – turbocharging being added later.

1969 saw the new 70 series – 470 and 570 – and that year also saw the introduction of the Agri-King line. These had fully enclosed cabs and included the four-wheel drive Model 1470, the largest tractor ever made by Case. In 1970 a family of four- and six-cylinder, in-line/open-chamber diesel engines were developed, producing

between 67 and 180 PTO horsepower. Case also started selling engines and hydraulic components to other manufacturers, and in 1972 the company experienced one of the most successful years in its entire history.

In 1982 the Panther 2000 tractor was introduced, the first model to have a 12-speed, full powershift transmission, electronic controls and PFC hydraulics. The following year a new line of 94 series, general-purpose tractors were announced, along with high-horsepower, two-wheel drive tractors, all in the 'power' red, black and white livery. That year the Super E was fitted with the newly designed Case four-cylinder diesel engine, which was produced at its facility in Rocky Mount, North Carolina.

Introduced in 1991, the Case 9280 is a massive machine. It uses a Cummins six-cylinder, turbo-diesel engine and has four-wheel drive with 12 forward and three reverse gears.

From the side, the 9280 looks big and weighs 13,154kg (29,000lb). It has a wheelbase of 3.6m (141.5in). The cost of this machine in 1995 was approximately US$13500.

The latest generation of the Case IH CVX range features five models between 137 and 196 horsepower. This is the 1170 model which produces 171hp.

In 1983 David Brown Tractors was renamed Case Tractors and integrated in the Agricultural Equipment Group. A year later Case introduced a new line of 94 series, four-wheel drive tractors, which included the most powerful Case tractor to date, the Model 4994, which boasted a turbocharged V8 engine.

In a move to increase its market position, Tenneco Inc. acquired selected assets from the International Harvester agricultural equipment

The Case IH, MXU MAXXUM X Line, a range of tractors that are straightforward to operate. The Surround Vision cab now features a lower profile for greater clearance.

operation in 1985, and the letters 'IH' were added to the corporate CASE logo.

A year later, Steiger Tractor Incorporated filed for bankruptcy protection, and Tenneco Inc. snapped them up. The new Magnum tractors were launched in 1987. The next year the new 9100 series Case IH Steiger tractors had Case's red livery.

In 1997 Case introduced its MX tractors and moved production to Racine, Wisconsin, and Doncaster, England. Agri-Logic, a leading software developer for agricultural applications, was acquired. Two years later, the Case Corporation merged with New Holland NV to create CNH Global. In 2000 the new STX series from Steiger was introduced, and Case covered the high-power market with its DX and CX ranges.

In 2013 they produced the Steiger 620, in both wheeled and Quadtrac tracked versions. Its maximum output of 692hp makes it one of the largest production tractors. The introduction in 2010 of the Efficient Power system saw it incorporated into Case's Puma, Maxxum, Magnum and Steiger tractors, helping to cement its place as a world leader in the agricultural industries.

This Quadtrac 600 from 2011 has a power rating of 600hp with a maximum output of 660hp. It was surpassed in 2013 by the 620's top power of 692hp, making it the world's most powerful production tractor.

Caterpillar

Three men were instrumental in forming one of today's best-known tractor companies. They each had thriving businesses before their merger, and they all had interests in the farming community. They were Benjamin Holt, Daniel Best, and C. L. Best.

Benjamin Holt was an American inventor who developed a design for one of the first practical Caterpillar tracks for use in tractors. He formed the Holt Manufacturing Company in the early 20th century.

Born in New Hampshire, USA, in 1849, Benjamin Holt was the youngest of eight brothers and sisters. The family owned a sawmill that processed hardwoods for the construction of wagons and coaches. Benjamin and Charles moved to San Francisco, where they started a new company in Stockton in 1883, doing business in hardwood, timber, and wagon-making materials.

During the Gold Rush of 1848, thousands of people moved to California, in the hope of finding their fortune. Many who failed started working the land, which was plentiful at that time. Local

FACT BOX

Caterpillar Inc. operates three principal lines of business

- **Machinery** includes the design, manufacture, marketing and sales of construction, mining and forestry machinery.
- **Engines** including the design, manufacture, marketing and sales of engines for Caterpillar machinery, electric power generation systems, on-highway vehicles and locomotives, marine, petroleum, construction, industrial, agricultural and other applications, and related parts.
- **Financial products** consist of Caterpillar Financial Services Corporation (Cat Financial), Caterpillar Insurance Holdings, Inc. (Cat Insurance), Caterpillar Power Ventures Corporation (Cat Power Ventures), and their respective subsidiaries. Caterpillar Logistics Services, Inc. (Cat Logistics) provides supply chain solutions and services to the company and over 50 companies throughout the world.

After much experimenting and tinkering, Benjamin Holt developed a track system. The original Holt machine, equipped with tracks rather than wheels, coped better on the soft soil than any wheeled vehicle.

Caterpillar

Seen here is a Holt Caterpillar being used to flatten bumps in the road. They were heavy machines and suited to this kind of work. Military work consisted of mainly towing artillery pieces.

These early, huge Caterpillar machines were ideal for clearing the ground and combating rough and dangerous terrain.

farms spread over thousands of acres and needed labour to manage their horses and harvest their crops.

Benjamin Holt bought up patents for farm equipment, worked on his own designs, and then increased the output of the Stockton Wheel Company. Their first combination harvester and thresher was hauled by 18 horses, and was sold in 1886. The large number of horses needed to pull these machines was a problem in itself, and so Benjamin set about finding a solution.

The first Holt steam-powered tractor came along in 1890, and although it was large and heavy, it could be powered by burning wood, coal or oil. It ran on wooden wheels and became very popular for harvesting large fields, its running costs being one-sixth of the horse-drawn method. There were some difficulties with manoeuvrability and the wheels would, on occasion, get bogged down in mud. Holt experimented with larger and wider wheels, but this only added to the problems. After a trip through the USA and Europe in 1903–1904 to study various ideas, including different track systems from a variety of people and manufacturers, new ideas followed.

The name Caterpillar was given in 1908 when a company photographer exclaimed, at the sight of it moving, that it crawled like a caterpillar. This is the Holt 45.

Daniel Best developed a portable grain cleaner, and later turned his attention to developing a combine harvester. The Best Manufacturing Company was formed in 1871.

It was about 1919 when the Best 60 tracklayer was produced. It weighed a hefty 9 tons plus, and used the company's own four-cylinder engine.

The first attempt at a track system came in November of 1904, when the rear wheels of a steam traction engine were replaced with a set of tracks Holt had just designed. Once in motion, the company photographer at the time commented that it crawled like a caterpillar, and in this way the Caterpillar name and trademark was born.

Petrol-powered tractors were developed during 1906, and the first model, Holt Caterpillar 40, went into production in 1908. By 1905 Benjamin's brothers had either moved on or died, and so he was left to run the company. The company was now known as the Holt Manufacturing Company and was producing tracked vehicles for farming, road construction and military vehicles.

Of all the tractor models Holt produced, the Caterpillar 75 was the most popular. But the other

The Caterpillar Thirty was produced in the 1920s. Although a Best machine, both Best and Holt models became Caterpillar after the merger in 1925.

33

Caterpillar

The Caterpillar RD6, later to become the D6, was manufactured in 1935. The RD6 Crawler Tractor was powered by a six-cylinder diesel engine of 78hp.

The Caterpillar Sixty was initially developed by the Holt Company, with rear drive tracks and a forward steering wheel. Holt tractors became successful in West Coast logging and farming industry.

"CATERPILLAR" SIXTY TRACTOR

A Caterpillar Sixty brochure. It had a familiar overhanging radiator, individually mounted cylinders, lever controls and open clutch. It was a heavy-duty work machine.

root of the future Caterpillar Corporation, which would be formed by Daniel Best and his son Leo Best, also has to be considered at this stage.

Daniel Best was born in Ohio, USA, on March 28, 1838, the ninth child of 16 from his father's two marriages. As a youngster, he lived for a time in Missouri, where his father also ran a sawmill. The family then moved to Vincennes, Iowa, to farm. One of his older brothers had already moved out west, and Daniel followed him in 1859. For the next 10 years he earned a living in a variety of ways, mostly connected with the mining and timber industries. While working with his brother, he designed and built a transportable machine for cleaning grain, able to clean up to 60 tons of grain per day. He patented it in 1871 and struck up a partnership with L. D. Brown, and the 'Brown and Best's unrivalled seed separator' won First Prize at the California State Fair. Best went on to patent a seed-

The Caterpillar Tractor Co. was formed on April 15, 1925, when the C. L. Best (depicted) Gas Traction Company of San Leandro merged with the Holt Manufacturing Company of Stockton, California.

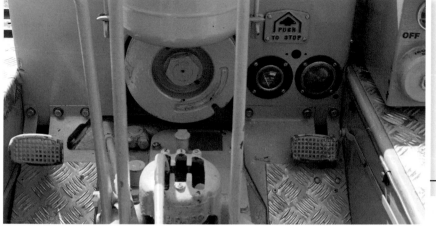

This is a driver's view of the controls for the diesel D2 Caterpillar crawler tractor. The gearshift is clearly visible, not that a great speed could ever be attained.

A good view of the four-cylinder diesel engine from the D2 crawler. It had a 3621cc (221 cubic inch) engine capacity and was made between 1938 and 1957.

coating machine and continued to work in the corn separator market. He next went into partnership with Sam Althouse of Albany, Oregon, and in 1879 opened a branch of their business in Oakland, California.

Best then moved with his family to Washington to pursue more mining and timber interests in 1879, while still a partner in the Oakland business. The demand for his inventions grew, and he started manufacturing a variety of machines aimed at increasing productivity and mechanizing farm work.

In 1882, while working for Nathaniel Slate, he was exposed to the workings of a combined harvester, giving him ideas for his own designs of this machine. Growth of the business made him decide to sell some of his other interests in Washington and Oregon, and he went on to buy the San Leandro Plough Company in 1886, renaming it the Daniel Best Agricultural Works.

During this period, Best patented a combined header and thresher and a fan blast governor that allowed the machine to work at a constant speed, regardless of what speed it moved across a field. Seen as a major step towards quality control in grain harvesting and cleaning, this also combined two functions in a single machine. Californian wheat farms were much bigger than others in the USA, and harvesting was hugely labour-intensive. Best was aware of the possibility of mechanizing

The D2, which weighed about 3,265kg (7,200lb), was initially manufactured in 1938, and upgraded in 1947 when it received a larger engine and more power.

Caterpillar

A scene showing a Caterpillar in its natural surroundings. This machine, fitted with a grader, is widening the road on the Alcan Highway during World War II.

the harvesting process to save both manpower and horsepower. Steam power was already being used in two forms for agricultural purposes: the horse-drawn steam engine as a source of power, and the self-propelled steam traction engine. Best bought the rights to manufacture the Remington 'Rough and Ready' steam traction engine, and redesigned it to tow his combined harvester and power its auxiliary engine. The successful machine was patented in 1889. In 1908, to the dismay of his son Leo Best, Daniel Best sold his tractor company to the Holt Manufacturing Company. Leo decided to start his own company, C. L. Best Gas Traction Company, in 1910, and within a few years he was building his own crawler.

During World War I the Holt company supplied the military with vehicles, while the Best company supplied the farm community: each had its area of operations. After World War I and throughout the early 1920s there was a tractor slump, and although the Holt business went through difficult times, Best saw his sales increased by

The first issue of Caterpillar Magazine was printed in December 1925. It was designed to showcase Caterpillar products and technology around the globe. This is Issue 34.

70 per cent. It was at this time that the two companies made a decision to merge, and although there were many factors that brought about the merger, it was not the idea of either the Holts or the Bests. Benjamin Holt died in 1920 and never saw the merger, and Daniel passed away in 1923. In 1925 the Holt Manufacturing Company and the C. L. Best Tractor Co. merged into the new Caterpillar Tractor Co. The merger worked well, and the products of the two companies complemented each other.

In 1927 a mid-size crawler, the Model Twenty, was introduced, which used an in-line, four-cylinder engine, and just one year later the Caterpillar Fifteen was introduced, with the first diesel engine tractor, the Diesel Sixty, being delivered to its buyer in 1931. This was also the first tractor to be painted in the now-familiar Hi-Way yellow livery, but it wasn't long before every model was painted this colour too.

During 1935 there were new crawlers with diesel engines – RD8, RD7, and RD6, with the RD4 introduced in 1936. Caterpillar, looking to fit

The R2 Caterpillar was produced between 1934 and 1942. It was revamped in 1938. It used a four-cylinder petrol engine of 4113cc (251 cubic inches) displacement.

The power unit from the D4 Caterpillar. This is the four-cylinder diesel unit, with a bore and stroke of 11.4 x 14cm (4.5 x 5.5in), and a displacement of 5735cc (350 cubic inches).

Made between 1938 and 1957, the D4 used a four-cylinder diesel engine. By 1934 Caterpillar were building more diesels than any other company in the world.

Caterpillar

First introduced in 1958, the D8H had a huge impact on the earthmoving industry. It had an operating weight in excess of 21,318kg (47,000lb) and a 235 HP turbocharged diesel engine for power.

CAT

CHALLENGER 65B

AGRICULTURAL TRACTOR

The Total Field Machine

■ Innovative Mobil-trac System (MTS) — the best of wheels and tracks.
■ Differential Steering — Automotive-type steering for precise, constant steering response.
■ Direct-Drive, Powershift Transmission — designed and built by Caterpillar specifically for agricultural tillage applications.
■ State-Of-The-Art Agricultural Cab — comfort and style, conveniently placed controls assure high productivity.
■ Power Applied Efficiently — converts more engine power to drawbar power than wheel tractors.
■ 30% Torque Rise — provides lugging power needed for heavy drawbar applications.
■ Total Customer Support System — parts or service..."when and where" you need them.

Cat® 3306 DITA diesel Engine
Gross power ....212.5 kW/285 HP
Drawbar power at 1900 RPM* ....168 kW/225 HP
PTO* ....186 kW/250 HP
Operating weight ....14 060 kg/31,000 lb
to 17 690 kg/39,000 lb

A page from a CAT brochure (with the new company logo) depicting the new heavy-duty Challenger 65B tracked tractor.

blades to their own machines, struck up relationships with at least six blade manufacturers, including LaPlant-Choate. The blade manufacturers produced their products to fit the crawlers, and Caterpillar encouraged their dealers to sell their products.

In 1938 the company introduced its smallest crawler, the D2, designed for agricultural work. The R2 variation followed, which could use paraffin or petrol power.

In 1942, when America joined World War II, track-type tractors, motor graders, generator sets and a special engine for the M4 tank were being supplied to the US military by Caterpillar. Once the war was over, production got back to normal, and pre-war models began to reappear alongside new machines.

In 1950 Caterpillar Tractor Co. Ltd set up office in the UK, the first of many overseas operations created to help manage foreign exchange shortages, tariffs and import controls, and to allow the company to better serve its customers around the world. In 1931 the company had created a separate engine sales group to market diesel engines to other equipment

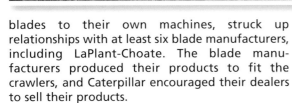

It is not often that you see a Caterpillar without tracks. Here is a rare view of one with high-flotation, low pressure tyres. It has obviously been well used.

The revolutionary track system of the Challenger tractor utilizes tough rubber tracks and an innovative suspension system, offering a comfortable ride and exceptional reliability.

The D9T cab interior is now larger and has a lowered beltline and low-profile door handles. There are also large, single-pane door windows with a wider opening door for better access.

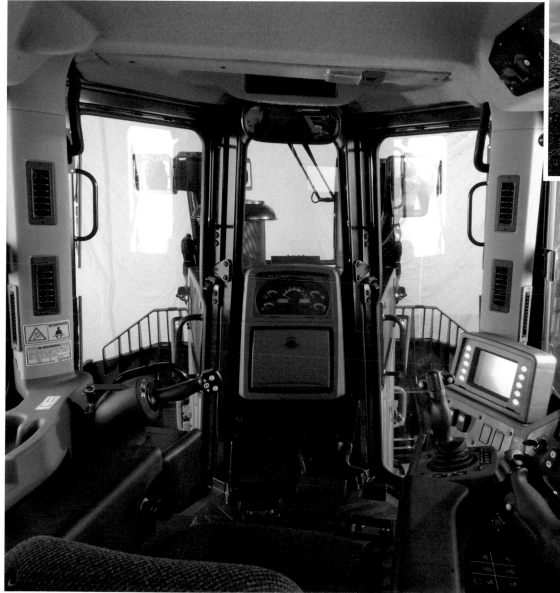

The D9T Caterpillar, introduced in 2004, was one of three new T-Series Track-Type Tractors. The machines were powered by Cat engines with ACERT Technology, which met strict exhaust emission laws.

manufacturers. This was replaced in 1953 with a separate sales and marketing division to better serve the needs of a broad range of engine customers.

In 1963 Mitsubishi Heavy Industries Ltd and Caterpillar formed one of the first joint ventures in Japan, which was also to include partial US ownership. Caterpillar Mitsubishi Ltd commenced production just two years later, with a company name change in 1987 to Shin Caterpillar Mitsubishi, becoming the second largest maker of construction and mining equipment in Japan.

By 1966 construction machinery was the company's biggest market; however, it hadn't forgotten the farming community, introducing the D4, D5 and D6 in SA (Special Application) form. The D3B SA began production in 1985, the D4D SA and D5 SA in 1966, the D6C SA in 1970, the D7G SA in 1977, and the D8L SA in 1984.

In 1976 the company introduced sealed and lubricated tracks to its vehicles, which, although reducing track wear, did not really help the farmers, who were looking for greater road hauling speeds and more comfort. This was remedied in 1987, when, after much development,

The D6R cab is designed for comfort and ease of operation. The cab is isolation-mounted and pressurized to reduce noise and vibration.

The D6R Caterpillar elevated sprocket undercarriage arrangements allow optimized balance for the best possible performance in each application.

After the 2002 sale of the Challenger brand to AGCO, Caterpillar no longer had any interests in the agricultural industry, concentrating instead on construction. In 2003, the company's 75th anniversary, they became the first manufacturer to offer a full line of clean diesel engines, certified by the US Environmental Protection Agency. The engines utilize the Caterpillar emission control system known as ACERT, with the second generation introduced in 2007 to meet stricter controls, representing a significant advance in the fight for a cleaner future.

Caterpillar produced its Mobil-trac system. Initially fitted to the Challenger 65 tractor, they were so successful that they were fitted to the other Challenger models too.

The recession that hit many industries in the early 1980s also had a huge effect on Caterpillar – at one point the company was losing the equivalent of one million dollars a day – and it had no option but to make dramatic reductions in the workforce. In 1986 the company changed its name to Caterpillar Inc., and 1987 saw the company spend some US$1.8 billion to modernize and streamline its plant.

The company began to expand once again, acquiring Germany's MaK Motoren in 1996, and UK-based Perkins Engines in 1998, consequently making it a world leader in the manufacture of diesel engines. Also that year, the world was introduced to the 797, a monster vehicle that took the title of the world's biggest off-road truck.

David Brown

David Brown, founded in 1860, originally made wooden gears for use in the textile mills of England. By the 1930s they had become the biggest gear manufacturer in the United Kingdom.

Harry Ferguson demonstrates his latest creation, the Ferguson Brown tractor to an expectant crowd. He made the tractor to complement his innovative attachments.

David Brown first became involved with tractors in 1936, as a subsidiary of the local family firm David Brown and Sons. It collaborated with Harry Ferguson in the manufacture of the Ferguson Brown Tractor.

Ferguson had designed a new three-point-hitch system and needed someone to build the tractor he had designed to go with it. David Brown agreed that Ferguson would sell the machine, and they would build it, initially at the Park Gear Works, Huddersfield, and later at the nearby Meltham Mills tractor factory. It had a four-cylinder, water-cooled, petrol or petrol/TVO engine. The first 500 tractors used a Coventry Climax-type E engine, while the remainder used a

FACT BOX

David Brown firsts

- **1937** World's first farm tractor equipped with hydraulic lift and converging three-point linkage.
- **1948** Two-speed power take-off.
- **1949** High-speed direct-injection diesel engine for farm tractors.
- **1953** Traction control (implement weight transfer).
- **1958/1966** Six-speed/ 12-speed gearboxes respectively.
- **1959** All-purpose tractor hydraulic system with single lever control.
- **1964** Dial-controlled tractor hydraulic system.
- **1968** Introduction of fully approved safety cabs for all models.
- **1971** Semi-automatic transmission providing on-the-move clutchless changes to any of four ratios in each working range.

This is the front end of a beautifully restored Ferguson Brown. The plaque on the front confirms that it was made by David Brown Tractors of Huddersfield, England.

David Brown

When Ferguson went over to America to demonstrate his three-point hitch to Henry Ford, David Brown decided to build his own tractor – this was the VAC 1.

The David Brown-designed VAC 1 tractor used a four-cylinder, overhead valve engine, which was manufactured by the company.

2010cc David Brown unit. It had three forward speeds and one reverse speed and independent wheel brakes. Sales were generally good, but tension set in following a downturn – Harry Ferguson was against changes that David Brown felt could improve the tractors. Approximately 1,350 of these improved models were built before Ferguson and Brown parted.

David Brown started developing its own tractor, and in September 1939 the new model was presented. The VAC 1 had elegant styling, sleek lines and proved an immediate success. It used a four-cylinder, water-cooled, petrol or petrol/TVO engine, and had four forward gears and one reverse gear. The track was adjustable by dished wheel centres – a David Brown patent – and implement depth was controlled by a patented depth (gauge) wheel system.

Just before World War II, David Brown purchased a redundant textile factory at Meltham Mills, Huddersfield, with the intention of building tractors there. They did for a while, but munitions and aircraft parts production took over. VAC 1 aircraft tugs and crawlers were also used for the war effort.

VAC 1 production continued after the war, and in 1945 the VAC 1A was introduced. This featured an improved engine lubrication system and better

The VAC 1 tractor was ready for production in 1939, but the war intervened. Although military versions were made during the war, general production started after the conflict was over.

governor. A patented automatic load-controlled hot spot, for rapid TVO engine warm-up, was introduced, and the turnbuckle top link – another David Brown feature – was also fitted.

In 1947 the Cropmaster was introduced, with its many standard accessories: hydraulic lift, swinging draw-bar and electric lighting. The two-speed PTO, six-speed gearbox, and coil ignition were all introduced by David Brown now too. The Cropmaster Diesel tractor was powered by the four-cylinder, water-cooled, direct ignition, cold-starting OHV engine – the cylinder block was a single piece casting fitted with four detachable wet sleeve liners of close-grained iron, and the camshaft bearings and main oil galleries were integral to the cylinder block. The long production run of Cropmaster tractors (1947 to 1953) saw many pioneering new features, and the machines greatly enhanced the reputation of David Brown. From 1953 to 1958 the 50D model was produced, a rugged, heavy tractor ideally suited for towing operations which featured a four-speed PTO unit. It was unique among David Brown tractors in having a side-mounted belt pulley. The 30C, a petrol/TVO model, and the 30D with its diesel

This is a typical VAC 1, modified for use by the RAF during World War II. It would have been used as a 'tug'. This involved moving aircraft around airfields.

The David Brown 25D was introduced in 1953, and was a great improvement on the petrol/TVO Cropmaster. It was considerably easier to start and had much better fuel consumption.

Made between 1954 and 1958, this is a David Brown 25 model. It used a four-cylinder diesel engine, and had six forward and two reverse gears.

David Brown

engine, had overhead valves and coil ignition, and by 1954 were equipped with TCU – Traction Control Unit. The 25C and 25D models, introduced in 1953, were the first small tractors to have TCU; they also featured a two-speed PTO, belt pulley unit and a six-speed gearbox.

In 1956 the 2D was introduced, which was ideally suited to precision market garden work, and as a specialist row-crop machine on larger farms. It used a lightweight, rear-mounted, air-cooled, two-cylinder diesel engine, with optional rear lift and PTO. Both lifts were operated by

The David Brown VAC 1C Cropmaster replaced the VAC 1A in 1947. It was the first tractor in the David Brown range to carry the Cropmaster name.

The David Brown 2D was produced between 1956 and 1961. Compressed air, supplied via a compressor unit driven by the tractor motor, allowed the two lift cylinders to raise and lower the under-slung tool bars.

The rear-mounted engine of the 2D was a 14hp, four-stroke, air-cooled, two-cylinder diesel unit. Drive was via a single-plate clutch, with four forward gears and one reverse gear.

The David Brown 40 TD crawler was not the biggest in the range, but it coped well with larger tasks. This one is being used for earth-moving duties.

The David Brown 900 tractor was presented in 1956. There was a choice of engines: 37hp paraffin, a 40hp or 45hp petrol unit, and a 40hp diesel version.

The David Brown 850 was made between 1960 and 1965. It used a four-cylinder diesel engine and had six forward gears and one reverse gear.

compressed air, with independently operated front lift cylinders. 1956 also saw the 900, available with four alternative engines, with the diesel model pioneering the use of a distributor-type fuel injection pump. In 1957 the 900 Livedrive was introduced – the first David Brown model with dual clutch. The 950 T and U series were similar in design but had increased power.

The introduction of the 950 Implematic in 1959 offered farmers the opportunity to use the depth (gauge) wheel system, or draught control with equal ease. In 1961 the V and W series were superseded by the 950 Implematic A and B series, which had improved front axle clearance and multi-speed PTO. The A and B series 850 Implematic tractors had a four-cylinder diesel engine, but petrol versions were also available. The later C and D series had diesel engines only, and featured multi-speed PTO and improved front axle clearance. From April 1963 height control was included in the hydraulic system, and September 1964 saw the 950 and the 880 Implematic models superseded by the E and F series 880 Implematic, with a new three-cylinder engine.

With the introduction of the 990 Implematic in 1961, the principle of a cross-flow cylinder head in conjunction with a two-stage front mounted air-cleaner was used. The 990 was powered by a 52hp direct-injection diesel engine, and in 1963 height control was introduced, the wheel-base was increased, a front-mounted battery fitted, and a

Along with standard tractors, David Brown also got involved with the manufacture of crawlers. Seen here hard at work is the 50 TD, one of the larger models.

David Brown

The 780 of 1967 – the Selectamatic hydraulic system (above) allowed the driver to select depth control, height control or external services on the Selectamatic dial.

The David Brown 885 was manufactured between 1971 and 1980. A cab could be fitted for protection against the elements, as seen on this machine.

12-speed alternative transmission was also fitted, along with a roll-over cage.

The 770 was powered by a three-cylinder 35hp diesel engine and had a patented two-lever, 12-speed Selectamatic gearbox as standard equipment. This proved so successful that it was introduced on the whole range in October 1965. The 770 was upgraded to 36hp and restyled in a new orchid-white and chocolate livery, and remained on sale until 1970. 1965 saw the 880 re-rated to 46hp and the 990 to 55hp, both equipped with multi-speed PTO, differential lock, Selectamatic hydraulic system and restyled in the white and chocolate livery. A four-wheel drive version of the 990 was introduced in 1970, along with full-flow filtration of hydraulic oil incorporated on all models. The 67hp 1200 Selectamatic tractor of 1967 – uprated to 72hp in 1968 – was the first David Brown model to have a separate hand

The David Brown Rose badge's roots are with the 13th-century War of the Roses. The Duke of York had a white rose emblem; the Earl of Lancaster a red rose.

The 885 model (below) had a choice of either a three-cylinder diesel engine or three-cylinder petrol engine, and weighed some 1,977kg (4,360lb).

The David Brown 990 Implematic, was introduced in the early 1960s. It was powered by a cross-flow, four-cylinder engine and rated at 52hp.

With David Brown now part of the Case IH corporation, many models were rebadged Case IH. This model 1594 still holds the David Brown name and badge.

clutch controlling the drive to the PTO. In addition, the hydraulic pump was mounted at the front of the engine and it had three-point linkage and luxury suspension seat. The four-wheel drive 1200 was announced in 1970 and the 72hp 1212 version followed.

In 1971 the Synchromesh 12 forward and four reverse gearbox became standard equipment on all tractors except the 1212. The 885 superseded both the 780 and 880, taking the best features from each, while the 990, now with Synchromesh gearbox, continued to be produced. The 996 featured a hand-operated PTO clutch, although, unlike the 1200 class, its hydraulic pump was retained behind the gearbox. The David Brown-designed safety feature, Weatherframe, was introduced in 1971, while the alternator became a standard fitting in 1973.

In 1972 the David Brown Tractor Company was sold to Tenneco Inc. of Houston, Texas, and became affiliated to another world-famous Tenneco subsidiary, the J. I. Case Company of Racine, Wisconsin, USA. 1988 saw the Meltham factory close, bringing to an end another significant piece of tractor history. Even so, under the Tenneco banner, David Brown Tractors and Case were now successfully expanding their combined production, marketing and distribution facilities.

Deutz

Deutz was founded by two pioneers of the internal combustion engine, Nikolaus August Otto and Eugene Langen. In the early 1860s they developed a four-stroke engine, and then formed Gasmotoren Fabrik Deutz AG.

Nikolaus August Otto (1832–1891) invented the four-stroke principle or Otto Engine. He met Eugene Langen, a technician, and in 1864 they established N.A. Otto & Cie., the first engine company in the world, and today known as DEUTZ AG, Köln.

DEUTZ FAHR

The two men went out of their way to employ the best German engineers of the time – men such as Gottlieb Daimler and Wilhelm Maybach – and in 1907 the Deutz company produced its first tractors, along with a motor plough. In 1926 the single-cylinder, water-cooled, MTZ 222 diesel tractor was unveiled.

The diesel engine had effectively changed the face of tractor manufacturing in Germany, and Deutz was at the forefront of its development for agricultural use.

At the start of the 1930s the company produced the diesel-powered Stahlschlepper (Iron Tractor) models, which included the single-cylinder F1M 414, the twin-cylinder F2M 312 and the three-cylinder F3M 315.

Soon Deutz were selling engines to other tractor makers, for example Ritscher, who used the engines in their tricycle-type tractor.

Both the F1L and F2L (shown here) were equipped with air-cooled diesel engines. Deutz was the first company in the world to offer a diesel tractor.

This Deutz D8 06 is hard at work in a field in the Netherlands. Although it was a reliable tractor, the engine got a reputation for being over-noisy.

With its Deutz turbo-diesel, six-cylinder engine, this is the Deutz-Allis 7145 model. The synchromesh gearbox gave 36 forward and 12 reverse gears.

Built between 1959 and 1964, the Deutz D15 used a ZF transmission with six forward and two reverse gears.

Manufactured between 1989 and 1992, the Deutz-Allis 9130 used a Deutz six-cylinder diesel engine with a capacity of 6128cc (374 cubic inches).

The Deutz tractor evolved throughout the early 1930s, and by 1937 the company made one of the first mass-produced mini-tractors, powered by a single-cylinder diesel engine, and producing 11hp. The company merged with the engineering firm Humboldt and engine-maker Oberursel, operating under the name Klockner-Humboldt-Deutz. They made a variety of vehicles during World War II, after which mass production was resumed, and Deutz not only made its 50,000th

tractor at that time, but also produced the best-selling tractors in France.

In the mid-1960s the company linked up with the implement maker Fahr, merging in 1969. In 1967 Deutz began exporting the D5506 and D8006 to the USA, even though air-cooled machines were not common there. By the mid 1980s Deutz had acquired Allis-Chalmers, and the first Deutz-Allis machines were a combination of German imports and surplus inventory of the Allis range. In 1986 the Deutz-Fahr DX range became the Deutz-Allis 6200 series; all were powered by Deutz air-cooled diesels. Rebadging then produced the Deutz 5015, 8010, 8030, 8050, 8070 and 4W-305. Over the next few years the Deutz content was increased, and in

After Deutz acquired the American Allis-Chalmers company, they mixed and matched their products and names. Seen here is the Deutz-Allis 6265.

The Deutz-Allis 9150 shown here is still painted in Allis orange. It was made between 1989 and 1992, and used a Deutz six-cylinder diesel engine.

1987 the 7000 series was imported from Germany, the air-cooled diesels offering up to 144hp. In 1989 the White-New Idea company built the 9130, 9150, 9170 and 9190 tractors for Deutz-Allis, all powered by Deutz engines ranging from 150hp to 193hp, and having 18-speed transmission with three-speed powershift.

Deutz-Allis was bought by the new AGCO corporation, and in 1992 the tractors were rebadged AGCO-Allis.

In Europe, the other arm of the old Deutz concern, Deutz-Fahr, was attracting the attention of a number of large corporations. Finally bought by the Italian group SAME in 1995, the company developed and manufactured a number of high-tech machines. The Agrotron, launched in 1995, was beautifully styled and used a water-cooled diesel engine designed by Deutz. Despite early performance glitches, its eventual success ensured its survival within the SAME fold. The Agroplus model, a much simpler concept modelled on the SAME Dorado, had a choice of Deutz engines between 60 and 95hp.

Deutz continue to operate under the SAME banner, winning the 2013 Tractor of the Year for the 7250 TTV Agrotron. Since its origins in the early 20th century, Deutz has become one of Europe's most durable tractor names.

The Agrotron models of the mid 1990s had a huge influence on future tractor design. Their raked engine housing allowing the driver to have excellent views ahead.

Fendt

When the Fendt brothers, Hermann, Xaver and Paul, of Weimar, Germany, started building tractors in a blacksmith's shop in the 1920s, they could hardly have foreseen the huge changes that would transform the industry in the 20th century.

The brothers Fendt introduced this 6hp engine, mounted plough and independently driven mower in 1930. Small and mid-size farms could now afford to replace horses with a tractor.

In 1928 Hermann Fendt built a mower – little more than a stationary engine with a transmission – mounted on wheels. In 1930 he and his brothers introduced a small tractor with a 6hp engine, mounted plough and independently driven mower. For the first time, small and middle-sized farming operations could afford to replace their horses. The name the Fendt brothers gave to their tractor programme, Dieselross, means 'diesel horse'.

On December 31, 1937, Xaver Fendt & Co. was established, and the following year the 1,000th Dieselross tractor, a 16hp F18, left the production line. Model F22 was introduced in 1938, with a two-cylinder engine, a conventional upright radiator and a four-speed gearbox. In 1942 shortage of diesel oil and the prohibition of diesel tractor operation led to the development of the methane generator tractor. In 1946 over 1,000 Dieselross machines were manufactured, and the company changed its engine supplier from Deutz to MWM – one of the last Deutz-powered Fendts being the 18hp F18H of

1950

Dieselross F15

This is the F15 model in the Dieselross series of tractors, of which there were many. The F15 model was presented in 1950, and used a 15hp diesel engine.

The Fendt Farmer 200 range is ideal for the small farmer. It is equipped with a three- or four-cylindet turbo engine. The narrow wheel base makes it ideal for vineyard work.

Fendt

1949. The 12hp Dieselross F12L was introduced in 1952, going into production in 1953. The Favorit 1, trend-setting in form, featured a 40hp engine, and the multi-gear close-ratio gearbox was produced in 1958.

In 1961 the 100,000th Fendt tractor, a 30hp Farmer 2, rolled off the production line. With about 60,000 models sold to date, the Tool Carrier, a mechanized system for a variety of tasks such as seeding and harvesting, reflected a particularly successful development. Another technical breakthrough came in 1968 with the Turbomatik, which was supplied with a stepless automatic transmission.

1952

Dieselross F12L

Another model from the Fendt Dieselross family is this F12L of 1952.

One of the most powerful machines available came in 1979 – the MAN-engined 262hp Favorit 626, followed in 1980 by the Farmer 300 range. In 1984 the 380 GTA System Tractor came on the market, and the following year Fendt took the market lead in Germany for the first time. High-tech and compact dimensions were the distinguishing features of the 200 range, produced as a standard tractor or as a special for wine and fruit growers. The 800 range was the first heavy-duty tractor with turboshift, hydro-pneumatic cab

In 1980 Fendt introduced the Farmer 300 range, featuring technological innovations such as a rubber-supported cab. This is the 309 LSA with Turbomatik stepless transmission.

Using a new 4 litre, four-cylinder, turbocharged engine, the 2002 300 model can tackle a multitude of farmyard duties. The 300 series offers tractors from 75 to 100hp.

and front-axle suspension. The Favorit 500 C-series (90–140hp) was introduced in 1984 and was capable of 50kph (31mph).

In 1995 the groundbreaking system vehicle Xylon (110–140hp) made its appearance. This vehicle could be used for agricultural, landscape and municipal applications, with a mid-mounted cab, and could carry front and rear implements simultaneously. With four-wheel drive, the Xylon

The trendsetting system vehicle Xylon (110 to 140hp) was introduced to the market in 1995. This was a universal vehicle for agricultural, landscape and municipal applications.

was powered by a turbo-intercooled MAN four-cylinder diesel engine, and the complex transmission gave 44 speeds. The Vario 926, produced in 1996, was the first heavy-duty tractor to have stepless Fendt Vario transmission.

In January 1997 Fendt became part of the corporate giant AGCO. In came the new Farmer 300C tractor generation, and the following year saw the new Favorit 700 Vario range. In 2002 the Favorit Vario was offered in two ranges with variable speeds, and remained the only stepless automatic system available on tractors of this power. A prototype of the 540hp, six-wheel-drive TriSix revealed in 2007 is yet to go into production.

Fendt has come a long way since Dieselross, and continues to be a solid European performer with a strong spirit of innovation.

The third generation of the popular Fendt 700 Vario was introduced in 2011, with an expanded output range of 145 to 240hp. It features the newly developed VisioPlus cab.

Ferguson

Henry George (Harry) Ferguson was born on the family farm at Growell, County Down, Ireland. In 1914 he started selling American tractors, which led him on to designing his own, along with his innovative three-point linkage.

A blue plaque on the Ulster Bank Building, Belfast, and granite memorial on the North Promenade, Newcastle, Ireland, are dedicated to a man who made the first powered flight in Ireland in 1909. Henry George Ferguson also made tractors.

Seen here is a Ferguson 35, using its front loader to load hay into a Ferguson trailer. The trailer is fitted with a rear-mounted manure-spreader.

Harry Ferguson, as he was known, was born on November 4, 1884, in the small Irish town of Growell, County Down, near Belfast. Even at an early age he showed an aptitude for all things mechanical, but, much to his father's dismay, had no desire to get involved with farming.

The young Ferguson had been fascinated by the exploits of the Wright Brothers and made a point of visiting air shows and exhibitions around Europe. On his return to Belfast, he persuaded his brother Joe that it would be good for their garage business to build a plane.

Construction took place throughout 1909, with various changes and improvements being made as work progressed. When the day of the first test flight arrived, the aircraft was towed along through the streets of Belfast, up to Hillsborough Park. Initial efforts to get off the ground failed, due to propeller trouble and bad weather. Finally, though, on December 31, 1909, the plane was ready. A reporter from the Belfast Telegraph described the scene:

"The roar of the eight cylinders was like the sound of a Gatling gun in action. The machine was set

An advertisement for the Ferguson System, which was very much ahead of its time. The system was envied by many of the other manufacturers, and was often copied.

Although he became more famous for his invention of the integrated tractor hydraulic system, Harry Ferguson made the first flight in Ireland in a plane he built himself, shown here.

With both the Press and local farmers watching, this is a demonstration of the Ferguson three-point hitch being carried out. The tractor is a Ferguson Brown.

FACT BOX

Ferguson timeline

- **1884** November 4 – Henry George Ferguson is born.
- **1909** Successful flight.
- **1913** Ferguson starts work on carburettor improvement.
- **1917** Belfast Plough – the first wheelless plough.
- **1925–1930** World's first automatic draft sensing three-point linkage system.
- **1933** Black tractor completed.
- **1936** Tractors produced by David Brown in Huddersfield.
- **1939** Ford handshake agreement.
- **1946** First TE 20s roll-off Standard Motor Co. production line at Banner Lane, Coventry, UK.
- **1952** April – Ford lawsuit settled.
- **1953** August – Merger between Massey-Harris and Ferguson.
- **1954** Ferguson sells out to Massey-Harris.
- **1960** October 25 – Ferguson dies.

The famous meeting between Harry Ferguson and Henry Ford, where they shook hands on a deal.

against the wind and all force being developed the splendid pull of the new propeller swept the big aeroplane along as Mr Ferguson advanced the lever. Presently, at the movement of the pedal, the aeroplane rose into the air at a height from nine to twelve feet, amidst the cheers of the onlookers."

Harry Ferguson had made the first flight in Ireland, and was the first Briton to build and fly his own plane. Assisted by the engineer Willie Sands, he went on to design the Belfast Plough, to be used with the Eros tractor, a conversion of the Ford Model T car. By 1917 the plough was ready for use as the first wheelless plough, and the first demonstration took place in Coleraine in front of the farming public.

About now, Ford had announced its intention to build a tractor plant in Cork, but unfortunately before the factory was ready, it decided to import 6,000 of its latest Fordson F tractors from America. Under-pricing the Eros, it sold in vast numbers, and

Ferguson

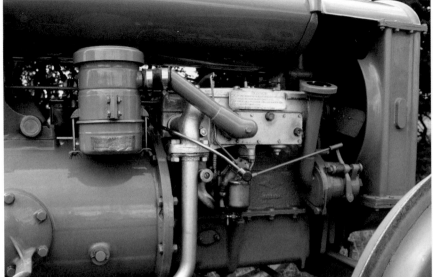

First produced in 1936, the Ferguson Brown tractor initially used a 20hp, side-valve, petrol/paraffin Coventry-Climax engine. Later versions were fitted with a David Brown unit.

This view from the rear of the Ferguson Brown tractor shows the steel spade lug wheels originally fitted. Within these are the independent drum brakes. It had no front brakes.

basically made the Ferguson plough redundant. Ferguson didn't dwell on his misfortune, though, and instead devoted his energy to designing a plough that could be attached to the Fordson F.

The result of this work was the famous Ferguson 2-point hitch, later referred to as the duplex hitch. Besides being comparatively light, it also prevented the tractor from toppling backwards if it hit an obstruction. This design established the basic principle of all subsequent Ferguson linkages, namely the concept of a 'virtual hitch-point'.

Seen here is the operator's seat, a rather uncomfortable metal dish. The gear-change lever is positioned between the driver's legs, and the hitch controls are positioned on the right of the operator.

The Ferguson FE 35 tractor introduced many improvements over the TE series, making it more user-friendly. Improved hydraulic lift system with duel lever control and a fine setting for the draught control lever were just two.

In 1938 the basic cost of the tractor was £198 – nearly 300 euros, and not much change from US$400. Compared to the £100 (150 euros/US$ 200) price of a Standard Fordson, farmers considered it to be too expensive.

Cutting the grass near a stately home in Somerset, England, this little grey Ferguson is right in its element. Note that the tractor is equipped with a roll-over bar immediately behind the operator.

In 1925 Ferguson came up with a second important invention – draft control. This enabled the depth of an implement to be automatically adjusted by reference to the effort needed to pull it through the soil. Further development finally culminated in the Ferguson three-point linkage.

Finding a manufacturer for these systems was hard going – the country was suffering from a deep depression and money was tight. Initially the Morris Motor Company showed an interest, but later pulled out. Ferguson decided to design his own tractor, which would accommodate all his new attachments. The prototype was assembled at his Belfast workshops during 1933, and many of the components, such as the Hercules engine and David Brown gearbox, were bought in. The tractor was

Ferguson

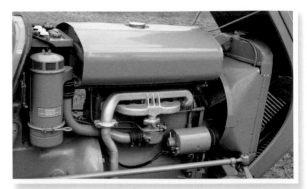

A fine example of a Ferguson TE 20 Continental Tractor, first of the much-loved Grey Fergies. Production started at the Standard Motor Company's Banner Lane factory in Coventry in 1946.

The TE 20 was powered by an imported overhead valve 24hp petrol engine, manufactured by the Continental Motor Co. in Michigan, USA. It had four forward gears and one reverse gear.

completed with Ferguson three-point linkage and draft control systems. A three-speed, constant-mesh gearbox took the drive to a spiral bevel rear axle, and independent brakes were fitted to assist turning. This became the infamous 'Black' Ferguson.

Ferguson now needed someone to build it in quantity, and this task was taken on by David Brown. The colour of the tractor was changed from black to battleship grey. In October 1938 Ferguson took tractor number 722, with implements, to the USA. A meeting was arranged for Harry Ferguson to demonstrate his tractor to Henry Ford. Ford was impressed, and the two concluded a deal with a handshake – agreeing that Ford would manufacture the tractor, incorporating all the latest Ferguson inventions and designs. After Ferguson returned to England Ford built two prototypes, but they proved entirely unsatisfactory and were discarded, so in February

The TE series of Ferguson tractors came in various designations. This is a 1953 TEF 20 diesel version. This same year Ferguson and Massey-Harris combined forces to become Massey-Harris-Ferguson.

The TO (Tractor Overseas) version was just slightly different from the others – note the dished wheels. This was basically an Americanized version that was built in Detroit.

1939 Harry Ferguson and a small team of engineers returned to Dearborn, USA, and under their supervision, work resumed on another new prototype, designated the 9N. Ford production engineers got the new tractor on line in record time, applying the very latest production line techniques. Ferguson had to accept the use of a Ford side-valve engine, and every effort was made to use stock items where possible.

When Henry Ford II took over the Ford empire in 1945, he wanted the agreement that had been made between Ferguson and Henry Ford terminated. This happened in 1947, and by 1948 Ford was building its own tractor. The whole affair ended up in court, and after long and costly proceedings, Ferguson accepted a settlement of $9,250,000, which only covered part of the claim.

During World War II, the Standard Motor Company had been using a factory making aero engines. This plant was in Banner Lane, Coventry,

A view from the 'business end' of this metal-wheeled, model TEA 20 Ferguson. This is an early 1948 example which has been lovingly restored.

Ferguson

and stood idle when the war ended. Standard's managing director, Sir John Black, and Harry Ferguson struck a deal, in which Ferguson would be in charge of design, development, sales and service, while Standard Motor Company would make the tractors. The TE 20 (TE standing for Tractor England) was produced, the first unit coming off the Banner Lane production line on July 6, 1946. It used an American Continental Z-120 engine, until the 2088cc unit of the Standard Vanguard car was fitted from July 1948, and a diesel version followed in 1951 – the TEF 20.

Ferguson also built a factory in Detroit, producing the TO20, an Americanized version TE, which was later uprated to the TO35, with more power and six-speed transmission.

When Massey-Harris merged with Ferguson in 1953, a prototype Ferguson 60 was developed, but never went into production. Once Harry Ferguson realized he was no longer in command of his company, he bought himself out.

The Ferguson TED 20, launched in 1949, differed from the TEA 20 in that it ran on TVO. It generally started with petrol, then switched to TVO once the engine was at temperature.

An advertising brochure for the Ferguson FE 35. In 1957 Ferguson terminated his interest in Massey-Harris-Ferguson. The grey and gold livery was dropped, and the tractor was renamed Massey Ferguson 35.

This beautiful model TEL 20 is clearly much thinner in its overall measurements than the other Ferguson tractors. This is a vineyard model, and runs on TVO (Tractor Vaporizing Oil).

Although the FE 35's optional four-cylinder, indirect diesel engine had a pre-heating system fitted, the series still gained a reputation for being difficult to start in cold weather.

Fiat

World War I was still being fought when, in a field outside Turin, a group of Italian industrialists and government officials met with businessman Giovanni Agnelli to view Italy's – and arguably Europe's – first agricultural tractor.

Italy, too, had its Empire, and many Italians made their homes in these far-off lands. Here a Fiat 700C is being used to clear the land in North Africa, where many Italians settled in the 1930s.

The Fiat Model 702 went on sale just after the conflict in 1919, and was made alongside the cars and trucks that the Fiat Turin factory was turning out. Built for ploughing and for powering stationary threshing machines the 702 became one of the most efficient tractors in the world, enjoyed very strong export sales, and played a major role in the mechanization of farm work that took place in the early 20th century.

The engine that powered the 702 came from one of Fiat's 3.5-ton trucks, a four-cylinder unit that produced 30hp in the petrol version, and 25hp when used with paraffin. Several other versions followed this, including an uprated 703 industrial model.

By the end of the 1920s, the smaller 700 series had made its debut in both wheeled and tracked versions, and these could also be fired up on diesel or petrol. The 700C crawler version of 1932 was probably the biggest selling Italian machine of the period and its success, along with the purchase of the Ansaldo, Ceriano SpA, and the OM, helped to strengthen the tractor division within Fiat. The crawler used a four-cylinder, 30hp engine and was equipped with a scraper blade on the front. In the 1930s the company also produced the 708C and the Model 40, the latter using a Boghetto engine that could be started on petrol and then switched to diesel once it reached working temperature.

During World War II the crawlers and tractors formed the basis of many machines used by the Italian military; the L6/40 tanks used SpA engines.

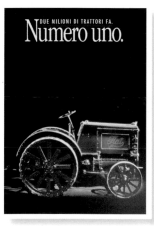

Clearly a best seller when introduced. This Fiat advertisement simply reads: "Two million tractors – Number one". This small machine was perfect for the small farmer, and was also not too expensive.

Giovanni Agnelli (1866–1945) was an Italian entrepreneur who founded Fiat car manufacturing in 1899.

This is the first tractor introduced by Fabbrica Italiana di Automobili Torino (FIAT). Work had started prior to 1919, but its presentation was delayed due to World War II.

Fiat

Probably a publicity picture of the tracked Fiat 700C crawler, taken near the factory. Note the wooden blocks fixed to the tracks. Their marks are also seen on the concrete floor.

Between 1971 and 1978 Fiat produced the 500 tractor model, of which this is the Special version. It used a three-cylinder Fiat engine, which had a displacement of just under 2621cc (160 cubic inches).

Fiat made a number of crawlers, for use both by the agricultural community and also for construction. They were rugged and tough workers.

Like many European countries, Italy was devastated by the war, and although able to sell many small cars after the conflict, it was hard going as far as tractor sales were concerned. With their tractors being large machines, they were too big and expensive for the average farmer of the period. Much like the automobile section, the solution came in the form of a tractor that could serve every kind of landowner. The Model 18, or *la piccola* (the small one), could cope with big jobs on small farms, and small jobs on large farms. It became extremely popular, with some 2,500 registered in the first year, with 30 different versions to choose from. The little machine had an 18hp engine with six forward and two reverse gears. Overseas sales were very important, and these too were buoyant in such countries as Romania, Yugoslavia, Turkey and Argentina.

In 1962 Fiat and the Turkish company Koa Holding set up a joint venture, by which time Fiat were offering a substantial range of vehicles, of all sizes and types. During 1966 the Italian company established a Tractor and Earthmoving Machinery Division, and in 1970 Fiat Macchine e Movimento Terra SpA was founded at Lecce, Italy, to pursue the company's activities in the earthmoving sector.

Although the first Fiat tractor, the 702, was made in 1919, it was still being manufactured up until 1924.

The Fiat 480 was produced between 1973 and 1984. It used a three-cylinder Fiat diesel engine and had a choice of either eight or six forward gears, and two reverse gears.

Simit, the leading Italian manufacturer of hydraulic excavators, was taken over, and in 1974 Fiat Macchine e Movimento Terra set up a joint venture with the American manufacturer Allis-Chalmers, creating the new corporation Fiat-Allis.

1974 also saw Fiat Trattori SpA being created, and in 1975 it became a shareholder of the Laverda concern. For 1977 Fiat took over Hesston in the USA, a move that allowed the company to directly enter the huge American market. At the same time it acquired Agrifull, which specialized in small to medium-sized tractors. A joint venture was set up with the Pakistani Tractor Corporation in 1983, and the following year Fiat Trattori became FiatAgri, the group's holding company for its agricultural machinery sector.

Between more mergers and acquisitions, all FiatAgri and Fiat-Allis activities were combined to create the new company FiatGeotech. Fiat then went on to acquire Ford New Holland, the name of the amalgamated corporations then becoming N. H. Geotech. Thus, through a complex process of integration, all the companies had come together under a common flag. The end of the 1980s saw the launch of the 90 series tractors, which ranged from 55 to 180hp.

The merger with New Holland had strengthened Fiat's grip on the larger modern tractor sector, and the New Holland 70 series

Fiat produced some very powerful and large crawlers. This is one from the Montagna series, the 505C. It used a Fiat model 8035, four-stroke, direct-injection, three-cylinder diesel engine.

This is the Fiat 780 model tractor, made between the mid-1970s and early 1980s. It used a four-cylinder diesel engine which produced 78hp.

The Fiat model 90-90 used a five-cylinder diesel engine which produced 90hp. It was produced between 1984 and 1991.

A huge machine, this 44-28 model had an articulated body and was powered by a Fiat turbo-diesel, six-cylinder engine which produced 280hp.

Produced in Turkey on behalf of Fiat, this is the model 80-66, which used a four-cylinder, direct-injection engine.

machines were now being sold in some markets as the Fiat G series – in Fiat red rather than New Holland blue; the name was merely part of a rebadging exercise that could work both ways.

The mid-1990s saw the introduction of the 66S series, ranging from 35 to 60hp, all using three-cylinder diesel engines. The larger Model 66S ranged from 65 to 80hp, and an even bigger 93 series was offered up to 85hp plus turbocharging.

Complex corporate acquisitions were far from over. When the Versatile Farm Equipment Company became part of Ford New Holland, N. H. Geotech's US division was further expanded. In 1993 N. H. Geotech changed its name to New Holland, marking the beginning of CNH. Another major corporate shuffle occurred in 2013 when CNH Industrial was formed through the merger of CNH Global and Fiat Industrial, which had demerged from the main Fiat company in 2011, taking all non-automobile interests with it.

This small-track model 70-66 is shown hard at work in muddy conditions. It was produced in four-wheel and two-wheel drive versions.

Ford

Henry Ford, best known for his major contribution to the automobile industry, was himself the son of a Michigan farmer, and was very aware of the need to apply the latest technology to farming.

Henry Ford was born on July 30, 1863. He grew up on the prosperous family farm in Dearborn, Michigan, USA. In 1879, at the age of 16, Ford left home and travelled to Detroit to work as an apprentice machinist. Little did he know what his future held.

The Ford 9N was produced between 1939 and 1941. The first of the N series tractors, it came complete with the Ferguson three-point hitch. Manufactured as an all-round tractor, it suited farmers with small farms, and became very popular.

In 1928, having achieved massive success with the Model F tractor, Ford closed down American production of Fordson, transferring operations to Ireland and the UK.

But Ford had not abandoned the idea of tractor production in the USA, and his engineers built and tested a number of prototypes throughout the 1930s. The eventual result was the Ford 9N, which used the Ferguson Brown tractor as its template, following a personal agreement between Ford and Harry Ferguson. The Ford 9N was revolutionary in its design. Low slung and relatively lightweight at 1,061kg (2,339lb), it was a new kind of utility tractor that could be applied to all kinds of farm work. But the real improvement was with the Ferguson System for implement attachment and control, which allowed the farmer increased speed and efficiency in the field. In 1942, however, Ford was forced to cease production of the 9N because materials bound for tractor production were diverted to the war effort. In its place, for a short period, came the 2N, a tractor designed to use materials that were not as scarce.

In April 1947 Henry Ford, the man who brought the motor car to the ordinary American, died at

The Ford 9N badge clearly shows the influence that Harry Ferguson had on the tractor. This was the end result of a handshake deal struck between the two men.

Ford

the age of 83. With his death, the agreement with Ferguson collapsed, and his grandson Henry Ford II announced that the company would distribute an improved version of the 9N, cutting out Ferguson entirely. This decision was to be a costly one, since Ferguson was now in a position to compete directly with Ford, but that was a few years away. In the meantime, Ford increased its market share with the 8N; this tractor would shortly become the company's best-selling model. One of the most important developments of the 8N was the four-speed transmission, although it still used the basic Ferguson System incorporated in the 9N line, and it was this unauthorized use that was the bone of contention in the lawsuit Ferguson filed against Ford. Despite high sales of the 8N, the lawsuit forced Ford to use a new hydraulic control system. This change was incorporated in the new Ford NAA, launched in 1953 and often called

A Ford advertisement for their Powermaster model, which is pulling a Ford combine. "More power per dollar – more performance per pound" was the slogan.

Ford

The 1958 Model 501 Offset Workmaster tractor is designed to carry underbody-mounted tillage tools, while also giving an unobstructed view. The Workmaster used the 2195cc (134 cubic inch) Red Tiger engine of the Jubilee 600 Series.

the Jubilee because its presentation marked the 50th anniversary of the Ford company. 1954 saw the introduction of the 600 and 800 series: the 600 was based on the NAA design and aimed at the small farm equipment market, whereas the 800 was aimed at larger farming applications. Ford was now interested in pursuing every sector of the tractor market. Accordingly, the 700 and the 900 were introduced, which were three-wheeled models, and could be used for a wide variety of applications, including row-crop work. In 1957 Ford decided to spruce up the appearance of its product line, and the '1' suffix was added in place of the '0' at the end of each model number.

Thus the 600 became the 601, and so on. In addition, the engine size on the 601 and 801 series was augmented by the addition of the Workmaster and Powermaster engines respectively. LPG was now also an option on all

The small Ford 4110 was produced between 1977 and 1983. It used a three-cylinder diesel engine and was aimed at the smaller duties on the farm.

Ford

The Fordson Super Major, produced at Dagenham in England, became known as the Ford 5000 in the United States. There were differences in the bodywork.

A Ford advertisement for their 66hp 6000 model, which was produced between 1961 and 1967. A six-cylinder petrol engine was fitted, but there was a diesel alternative.

Ford tractors. In 1957, Ford began to feel that the lack of an efficient small tractor was hurting its own position in Europe. To address this problem, Ford developed the Dexta, powered by a Perkins three-cylinder diesel engine. In 1958 the Fordson New Major was replaced by the Power Major, which used a more powerful version of the New Major engine. When the Super Major replaced the Power Major in 1961, the Super Dexta was produced to replace the Dexta. In 1959 Ford had introduced the Select-O-Speed transmission system, which provided 10 forward and two reverse gears, designed to give farmers greater control over rough terrain. In late 1961 Ford introduced the 2000 series to replace the 601 line, the 4000 to replace the 801 series, and the 6000, with a powerful six-cylinder engine, as the top of the range.

The model 6000 took Ford into the more powerful row-crop sector. The Select-O-Speed system, derived from and experimental tractor, often gave problems, however.

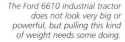

Totally the wrong colours – you could be forgiven for thinking this was not a Ford. In fact, it is the model 334 of 1982.

The Ford 6610 industrial tractor does not look very big or powerful, but pulling this kind of weight needs some doing.

Meanwhile, in a move towards unification of the US and European tractor arms of Ford, the Fordson Super Dexta was imported from European plants and sold in the USA as the Ford 2000 Diesel, while the Fordson Super Major was brought in as the Ford 5000. The Fordson badge was dropped and all Ford tractors now bore the same badge and came in the same blue-and-grey colour scheme. The world tractor line that Henry Ford had always favoured was now a reality. In 1965 the range from the 2000 to the 4000 was revamped with a new three-cylinder diesel engine. The 5000 was equipped with a four-cylinder diesel, while the 6000 was renamed the Commander 6000 and redesigned to address the model's earlier technical problems.

In the 1970s Ford turned its attention towards mini tractors. So successful were they that by the end of that decade the company was offering an entire range of them, from 11hp to 27hp. All Ford's smaller tractors now bore the '10' badge, which denoted three-cylinder diesels for most of the mini-tractors, and a 16-speed option for the mid-range machines.

Not yet in Ford colours, this is the Versatile 276. The Versatile Farm equipment company was taken over by Ford in 1987.

The Ford six-cylinder, turbo-diesel TW-15 was produced between 1983 and 1990. Here, one is seen helping to collect grain during the harvest season.

Although the Ford logo is clearly marked on this machine, it was also known as the New Holland 8210, Ford having taken the company over in 1985.

In 1985 Ford bought out the implements manufacturer New Holland, so now the company had a full line-up of farm tools to go with its tractors. But not long afterwards Ford announced its intention to pull out of tractor construction in the USA altogether. Production of smaller machines would be transferred to Basildon in the UK, while larger ones would be based in Antwerp, Belgium.

In 1987 Ford bought out the Versatile Farm Equipment Company, and a range of super-tractors was introduced by what was now known as Ford-New Holland Inc. Versatile's existing machines became Ford Versatiles in the Ford blue-grey colour scheme, and were powered by Cummins V8 engines. Ford introduced the 20 series in the late 1980s. Nine-speed transmission was included on models from the 1120 to the 1520, while front-wheel assist and hydrostatic transmissions were optional. In 1990 the mid-range 30 series was introduced. The four smaller models all had three-cylinder diesels and eight-speed transmission. By contrast, the 8530, 8630, 8730 and 8830 had 16-speed transmission and replaced the big two-wheel-driven TW series.

In the early 1990s the end of Ford tractor production was nearing. In 1991 Ford sold 80 per cent of Ford-New Holland-Versatile to Fiat, and three years later the remaining share was also bought by Fiat.

This was originally a Versatile, but then became the Ford 946. A considerable amount of badge engineering had taken place, and there would be more to come.

One could be forgiven for any confusion. Ford/New Holland/Versatile were bought out by Fiat Geotech. Seen here is the Ford-New Holland 8340, a top-of-the-range model.

Fordson

Henry Ford, the son of a farmer himself, began experimenting with tractor designs as early as 1907. However, he found it impossible to interest the directors of the Ford automobile company in the agricultural sector.

As early as 1906, Henry Ford had started looking at ways in which he could replace horse power on his farm. By 1914, experimental Model T type tractors were seen working the land around the Ford Farms.

In 1917 Ford was forced to set up another company – Ford & Son Inc., later simply Fordson – to market the tractor he finally developed. The Model F Fordson (the product, it is said, of up to 50 different prototypes) represented a great leap forward in tractor design. The engine and gearbox were taken from a Ford Model B car, and the steering adapted from a Model K. But unlike most tractors of the time, the engine, transmission and axle housings were all bolted together to form the basic structure of the machine. What was more, the tractor was comparatively light at 1,229kg (2,709lb), and a small 4.1 litre engine was sufficient to power it.

FACT BOX

Fordson models and production

- **Fordson Model F**
  1917–1928 Dearborn, Michigan.
  1919–1922 Cork, Ireland.

- **Fordson Model N**
  1929–1932 Cork, Ireland.
  1933–1945 Dagenham, England.

- **Fordson All-Around**
  (also called Fordson Row Crop)
  1937 Dagenham, England.

- **Fordson Major E27N**
  1945–1951 Dagenham, England.

- **Fordson New Major**
  1952–1958 Dagenham, England.

- **Fordson Dexta**
  1957–1961 Dagenham, England.

- **Fordson Power Major**
  1958–1961 Dagenham, England.

- **Fordson Super Major**
  (called the Ford 5000 in USA)
  1961–1964 Dagenham, England.

- **Fordson Super Dexta**
  (called the Ford 2000 Diesel in USA)
  1962–1964 Dagenham, England.

This is an early photograph dating from around 1906, showing Henry Ford with just one of his many experimental tractors, based on the Model T car. It is pulling a disc harrow.

Fordson

After several experimental tractors – none that were put into production – Henry Ford and son Edsel came up with the Fordson Model F in 1915.

Using the casings of the engine and transmission as a supporting frame, along with materials developed and used in the car industry, Ford created a lightweight tractor, the Model F.

At first the Fordson Model F was produced on a limited scale, but in 1918 production was increased. The ravages of World War I meant that Britain was suffering from a shortage of men, horses and grain. Tractors and mechanized farming methods seemed to be the solution. Accordingly, following highly successful trials of the prototypes in the UK, the British government ordered 6,000 Model Fs. In order to make machines on such a scale, Ford had to resort to the assembly-line methods he had developed for the Model T car. In the event, the Model F's compact frame made it perfect for this kind of quick, cheap manufacture. Again, as with the Model T, cheap production meant the machine could be sold at a price the average farmer could afford. Mechanized farming for the masses had truly arrived, with many Model Fs also being exported to Russia, while many thousands more were made there under license. By 1927, it was estimated that over 70 per cent of all tractors in the country were Fordsons.

In the USA, the Model F was selling in huge numbers. Rival tractor manufacturers were finding it increasingly difficult to compete with the low price of the tractor, and many companies went out of business as a consequence. Although sturdy and manoeuvrable, the Model F was not the perfect tractor. It often suffered from wheel-slip, and its

A fine example of a Model F, beautifully restored and in perfect working order. The Model F must have been a farmer's dream come true, and it went on to be a very successful tractor.

The engine of the Model F was a four-cylinder, side-valve unit, which produced 20hp at 1,000rpm. The ignition system was similar to that used by the Ford Model T car.

The initial Model F tractors had no guards over the wheels, which must have seemed daunting for the operator, especially with the very rough terrain these vehicles had to work in.

draw-bar performance was only adequate, while it also had a tendency to turn over on rough terrain, making it highly dangerous to the exposed driver. Generally, the Model F was reliable, cheap to buy and cheap to run, and by 1928 some 750,000 of them had been produced. That year, however, Ford decided to stop making Fordson in the USA – his Model A car was almost ready, and he wanted to devote all available factory space to its manufacture. Production of the Model F was transferred to a new factory in Cork in Ireland, where the model was transformed into the Model N. The Model N was essentially an updated Model F, although it was more powerful, had stronger front wheels and better wheel-grip. It also had a governor, a water pump, and a high-tension magneto that replaced the Model F's rudimentary ignition. However, the Irish relocation was not a great success, as Ireland, still a relatively isolated place, was cut off from Fordson's two main markets – the USA and the UK. Furthermore, all raw materials had to be imported, which pushed up production costs.

In 1933 the Fordson plant was moved again, this time to be incorporated in the new Ford complex at Dagenham in England. (Fordson, however, remained a separate arm of the Ford business.) The Model N did not undergo many changes, although the colour was changed from grey to blue. The tractor sold well in the UK, but Fordson's sales in the USA now took a downturn. It seemed that

Fordson

Ford's competitors, such as John Deere, Farmall and Allis-Chalmers, had managed to develop machines that could equal and out-perform the old Model F and the new Model N. Fordson's response was the All-Around, a general-purpose row-crop tractor. But this also failed to impress the American market.

In 1939 the story of Ford tractors – which comprised both Fordson in England and the Ford tractor arm in the USA – effectively divided. The Dagenham arm chose not to build the new 9N developed in the USA, preferring to continue with

Initial production of the Model N was in Cork, Ireland, but in 1933 production was moved to Dagenham, England, where the tractors were painted orange.

The Model N Standard Fordson was fitted with a 27hp engine. A high-tension ignition system, water pump and mudguards were standard equipment.

The Model F was followed by the Model N, the tooling for which was set up in Cork, Ireland. It had major differences, such as a larger horsepower engine. Production started in 1929.

The Fordson Major E27N was manufactured at Dagenham, England, between 1945 and 1952. It used either a Perkins six-cylinder diesel or a Ford four-cylinder kerosene engine.

An odd-looking machine, this is the row-crop version of the Model N. This model sold reasonably well on the British market but did not do so well in the United States.

the old model, which had acquired a loyal following in the UK.

In 1945, when the US factory was mass-producing the successful 9N, Dagenham chose simply to produce an update on the Model N theme – the E27N. The tractor was powered by an in-line, four-cylinder, side-valve engine that produced 30hp at 1450rpm. It came in four versions, each with different specifications for brakes, tyres and gear ratios. A variant using a Perkins engine was offered in 1948, and in that year over 50,000 E27Ns were made. Production continued until 1951, with various upgrades and options – including electrics, hydraulics and diesel engines. A number of E27Ns were also imported by Ford in the USA and sold there as the Major. The Major was a bit of a novelty on the US market, having both the diesel option and the high clearance that many US-built tractors lacked.

In 1952 the Dagenham plant replaced the E27N with the New Major. The New Major was larger

Clearly seen on the front of this Model N are the four rings of the Perkins diesel engine company. The diesel engine made this tractor a much better machine altogether.

It is not often that you see a Fordson Major with a cab. Specifically designed for the tractor, this was a delight for the operator during the winter periods.

The Fordson Dexta 957E tractor was launched in 1957. It was a completely new design and was fitted with a 30.5hp, three-cylinder, direct-injection diesel engine.

and heavier than its forerunner, and its diesel, petrol or TVO versions all used the same cylinder block and crank. All three versions had six-speed transmission and hydraulic three-point-hitch, although draft control was absent. The 1958 Power Major rectified this lack, increasing power to 43hp, and making power steering an optional extra.

The next tractors to come out of the Dagenham plant were the Super Major and the New Performance Super Major. Both of these models had differential lock, disc brakes and, finally, draft control. Later on, the Super Major would be sold in the USA as the Ford 5000, while the smaller Fordson Dexta, which had 31hp in diesel form, also became available on the American market. The Dexta and Super Major were the last tractors to bear the Fordson name. In 1961 Ford combined its US and UK tractor operations, and the two ranges were also fully integrated.

A well-restored Fordson Major – this is the diesel version. In the 1950s diesels were still not that reliable, and people still worried about their durability.

Hurlimann

Hurlimann produced its first tractor, the 1K, in 1929 and was one of several Swiss tractor companies to start trading during that period. The machine used a single-cylinder engine producing 10hp.

With Hurlimann being a branch of the SAME group, this well-restored 1951 Hurlimann H 12 is kept in the group museum. It used a four-cylinder petrol engine.

The D100 was presented in 1946, the first diesel tractor to use direct fuel injection, its Hurlimann water-cooled, four-cylinder engine producing 45hp. The H12 was introduced in 1951, and it used a four-cylinder, liquid-cooled, petrol engine. It was equipped with rear hydraulic lift and double towing hook for accessory attachment. This large-capacity machine had five forward gears and one reverse gear, and could reach a steady 20kph (12.4mph). The multi-purpose D90 tractor, satisfying the toughest of

FACT BOX

Hurlimann models and production

- **1929** First Hurlimann tractor using single-cylinder gasoline engine.
- **1939** 4DT45 tractor is presented, with the 4DB85 engine – first four-cylinder diesel tractor with direct injection.
- **1940** Hurlimann diesel tractors modified to run on carbon gas.
- **1950** D600 tractor is presented, exported mainly to Argentina.
- **1958** D90 is produced in Switzerland, the brand's most popular model.
- **1972** D115T tractor production starts.
- **1977** Hurlimann merge into the SAME group. Hurlimann T6200 tractor is presented – 62hp engine, four-wheel drive and central transmission (SAME system).
- **1979** The new 'H' series range of tractors, from 60 to 160hp, are presented, equipped with H-1103, H-1104 and H-1106 engines.
- **1991** The MASTER models are launched – equipped with Electronic Power Shift transmissions.
- **1993** PRINCE small tractors are presented.
- **1995** XA607 tractor with two- and four-wheel drive announced.
- **2004** XT130 presented, using 1000.4 WT engine and 60 forward and 60 reverse gears. XE F Tradition presented, using 1000.3 WT engine.

The 1930 Hurlimann 1K 10 is seen here with grass-cutting attachment. The engine is a single-cylinder 10hp unit, and it has three forward gears and one reverse gear.

Hurlimann

As shown in the bottom left of this 1979 advertisement, the company was celebrating 50 years of production when they introduced the H-490 4 RM.

The Hurlimann H-480 of 1979 was equipped with a new Hurlimann 82hp four-cylinder, liquid-cooled engine. There were both two- and four-wheel drive versions.

H-490 4 RM

2003 saw the introduction of the XT130. Seventy-two forward and 72 reverse gears worked in tandem with the new EURO II, 1000 series, six-cylinder engine.

The XB Max range of 85–110hp are versatile all-purpose tractors, ideal for use in small farms, and were given the latest four-cylinder emission-compliant Deutz engine in 2012. The Deutz six-cylinder, turbocharged engine is used in Hurlimann's current top-rated series, the 130–185hp XL machines. At the other end of the power range are the specialist XS compacts. The small dimensions allow easy access to narrow rows for specialist fruit growers, and the Euro II engine ensures plenty of power at the PTO. These are just a few of the exciting models in the Hurlimann catalogue for 2013.

requirements, was introduced in 1959, equipped with a double clutch (transmission and power take-off), ten gears and double hoist.

The T6200 was introduced in 1976. This had new styling, a four-cylinder Hurlimann 62hp engine and partially synchronized gears. Between 1977 and 1985 came the two/four-wheel drive H480, using a new Swiss-designed engine with separate cylinder heads. The cab had a platform which was mounted on silent-block, with ventilation and heating.

Since 1977 Hurlimann have been a part of the SAME group, and have developed a good array of models since then, including the XT 115 and 130 tractors fitted with the Euro II, 1000 Series six-cylinder, liquid-cooled, turbocharged engines.

International Harvester

The early history and background to the formation of the giant International Harvester concern is so important that to start from when they first built tractors would be to miss out an important period of the company's history. We need to look at the families behind the company.

Our story starts with the birth of Cyrus Hall McCormick on February 15, 1809, in Rockbridge County, Virginia, USA. By 1031, and now aged 22, Cyrus had taken over the design of his father's rather unsuccessful reaper, and within six weeks he had modified, built, tested and remodelled the design, down at the family smithy at Walnut Grove. The reaper was successfully demonstrated on a neighbour's farm at Steeles Tavern in late July of that year, and after making various design changes, Cyrus patented the machine in 1834. During the 1840s, Cyrus and his family manufactured and sold reapers from the blacksmith shop at Walnut Grove.

Cyrus Hall McCormick invented the reaper in 1831 and was instrumental in the creation of International Harvester. IH were swallowed up by Case, but the name McCormick still lives on in the tractor world.

In 1847 Cyrus moved to Chicago, where he built a factory; this Michigan Avenue site would later become the International Harvester corporate headquarters.

Disaster struck in 1871 with the Great Chicago Fire, which destroyed the McCormick reaper works, but fortunately the company safe was retrieved with all records intact. In August of the

Although the Mogul was an International Harvester model, it carried its own name, along with the company name.

International Harvester

The McCormick reaping machine was tested at Steel's Tavern in 1831. This is an illustration showing the slightly bemused crowd watching on.

following year, work was started on a new factory, which was completed in February 1873.

On May 13, 1884, Cyrus Hall McCormick died. He was survived by his wife Nettie Fowler McCormick and his brother Leander, who had joined him in partnership in 1856. During the late 1800s McCormick found itself increasingly challenged by competitors in a somewhat depressed market, its chief rival being the Deering Harvester Company. William Deering had established a rival harvester factory at Plano, Illinois, which he decided to move to Chicago in 1880. During the late 1890s, weary of competition, the Deering and McCormick families began to talk about a merger of their companies. So it was that in 1902, McCormick and Deering merged to form International Harvester. The Plano Manufacturing Company, suppliers of harvesting

Initially, the William Deering company was a great rival to the McCormick company, but the two finally got together and created International Harvester in 1902.

International Harvester used a number of brand names to market their products, including Mogul (above) and Titan.

The Mogul 12-25 was presented in 1913. It was an opposed, two-cylinder machine, with a disk plate clutch, and chain final drive on both sides of the tractor.

This is a close-up view of the pulley attached to the Titan 10-20. A leather belt could be attached, allowing it to drive threshing machines and the like.

1914 saw the introduction of the 8–16hp, single-cylinder Mogul tractor. This was a two-plough machine, with final chain drive, and differential mounted on the back axle.

Probably one of the most popular and versatile tractors of its period, this Titan 10-20 model was in production from 1914 to 1922, with some 78,363 being produced.

machinery, became the third part of the newly formed International Harvester organization, and completing the amalgamation were the Milwaukee Harvester Company and the Warder, Bushnell & Glessner Company. Cyrus H. McCormick Junior was named president, and Charles Deering became the chairman of the board. In 1903 International Harvester built a plant in Hamilton, Ontario, Canada, and two years later it built its first plant in Europe, at Norrkoping in Sweden, to manufacture implements.

The company unveiled its first production tractor in 1906, powered by an International 'Famous' single-cylinder engine. It was available in 10, 12, 15 and 20hp capacity, and was soon followed by a second model, the single-cylinder Type A, in late 1907. In December 1906 International Harvester of Great Britain (IHGB) was inaugurated, initially with offices in London, subsequently moving to Finsbury Pavement, where it remained until 1926. A new site was also purchased in Neuss on Rhine, Germany, during 1908, and by 1911 production of implements under the McCormick name had started. IHC expansion in Europe continued in 1909 with the building of factories at Croix, northern France and Lubertzy, Russia.

After the merger of McCormick and Deering, US dealerships were quite insistent regarding continuing to sell separate products, which caused

Seen here is the McCormick-Deering 10-20 model, which was introduced in 1923 and remained in production through to 1939. Over 200,000 were built during its lifespan.

An interesting feature of the Junior was that it had its radiator mounted behind the engine, and therefore was protected from the dust and dirt churned up by the tractor when working.

some concern. The government dictated that even though these companies were now under the IHC umbrella, they should be able to sell separate tractors and equipment. The conclusion was that McCormick dealerships would sell the new Mogul tractors, and the Deering dealerships would sell the new Titan tractors.

The Type C Mogul appeared in 1909, with a 25hp model presented two years later. This was followed by the 45hp model, using a twin, opposed-cylinder engine, and later upgraded to become the 30-60. A single-cylinder Mogul junior came in 1911, and then a Mogul 12-25 for 1912, a relative lightweight compared to the others.

In 1910 the Type D single-cylinder Titan was introduced, initially as a 20hp machine but later increased to 25hp. Other models appeared in the following years; 1911 saw the 45, later to be redesignated 30-60, and by 1915 it had been given a cab. Most of these Titan machines were built at the IHC Milwaukee factory, whereas the Mogul was made at the Chicago factory. The 1915 Titan 12-25 was upgraded as the 15-30 for 1916, and then became the International Titan 15-30. With some 14,000 units sold, the Mogul 8-16 was a big success and was in production until 1917, when it was replaced by a similar model, the 10-20. This machine used a single-cylinder engine, with two forward speeds, and mudguards for the rear wheels.

The Titan brand was first introduced in 1910, and continued until 1924.

The International Junior 8-16 kerosene tractor was produced between 1917 and 1922. The engine was an advanced overhead valve, four-cylinder unit.

The Farmall tractor started life in 1924 with limited production. It grew to be the most popular tractor of its time. This is the larger F-30 model, which was introduced in 1931.

The McCormick-Deering W30 was in production from 1932 to 1940, and was rated as a three-plough tractor. It used a four-cylinder, overhead valve engine and had four forward gears and one reverse gear.

FACT BOX

Farmall tractor

- First general-purpose tractor with narrow-spaced front wheels.
- Introduced by McCormick-Deering division of International Harvester (IH) in 1924.
- Initially there was only one model of the Farmall, but IH later developed more powerful models.
- The first Farmall model subsequently became known as the 'Regular' to distinguish it from later models which carried F designations.
- These were followed by the 'letter series' tractors (A, B, C, H, M), designed by Raymond Loewy.
- The Farmall H, produced from 1939 to 1952, became the top-selling individual tractor model of all time in North America, with over 390,000 sold.
- The Farmall Cub A and B models have the seat offset from the engine, allowing the operator to look directly at the ground under the tractor. This feature was dubbed Culti-Vision.

- The first Farmall tractor with an optional diesel engine is the M. It started on gasoline and was manually switched to diesel after warming up.
- The next Farmall tractor to offer diesel power was the 350, using a 350 Continental Motors engine.
- The Torque Amplifier (TA) was first introduced on the Super-M in 1954.
- Case IH has revived the Farmall brand on some of their latest tractors.
- A brief family tree of Farmall tractors based on number of ploughs:

1-plough: A, Super-A, B, BN, 100, 130, 140.

2-plough (sm): F-12, F-14, C, Super-C, 200, 230, 240, 404.

2-plough (lg): Regular, F-20, H, Super-H, 300, 350, 340, 504.

3-plough: F-30, M, Super-M, 400, 450, 460.

4-plough and up: 560, 656, 706, 806, 1206.

In 1917 things changed once more. The US Justice Department noted that IHC had too many Deering and McCormick dealerships, contravening competition laws, and so ordered that they should be consolidated. So from now onwards, each sales territory would have one dealership to sell one brand of tractor, and that would now be an International. So the International 8-16 was born, a radically different machine to its predecessors, using the company's own four-cylinder engine. This 8-16 tractor was a significant step forward as far as tractor design was concerned, and it helped to set them on the path to lighter and more economical machines. In 1921 International introduced their 15-30, in an attempt to counteract the onslaught from the Ford Model F.

The Farmall was the tractor that made the International name, and gave it the financial backing it needed. The F-14 model was derived from the F-12 and was introduced in 1938.

International Harvester

International Harvester had started experimenting with crawlers for a while before introducing the T20 TracTracTor. Production continued into 1939.

The Farmall M, one of the larger tractors in the letter series, is capable of pulling a three- or four-bottom plough. Production began in 1939 and continued until 1952.

A lightweight machine using the company's own four-cylinder engine, the 15-30 was also the first IHC tractor to use a unit frame design. Although not as cheap as the Ford, some 128,000 were delivered to customers over an eight-year period. A new 10-20 model came in 1923, again with a four-cylinder engine, and was produced up to 1939. IHC were making good progress with their tractors but it was their next design that help them to become market leaders.

Farmers were somewhat restricted in what they could achieve with their tractors. There were currently two types: the heavyweight used for belt work and the smaller machines, used for cultivating row crops. The Farmall, introduced in 1924, was capable of both jobs and more, but was initially only put on sale in limited numbers, for fear that it would stifle sales of the 10-20. The Farmall used a four-cylinder engine and had a good selection of attachments, helping it to be a truly multi-purpose tractor. The Farmall F-20, presented in 1932, was built up to 1939 and sold 149,000 units during that period. The more powerful F-30 version came in 1931, still with a four-cylinder engine, but with increased bore and stroke. Although it still used steel wheels, later in

The first International Harvester tractor to use a six-cylinder engine was the standard tread W-40. It was made between 1934 and 1940, with some 6,500 units being produced.

BASIC STANDARD NUMBER ONE

Internationally renowned industrial designer Raymond Loewy was invited to give not only the tractors but also the sales outlets a new and fresher look

Seen here rolling across the ploughed field is a TD-8 International crawler. This is the diesel version of the medium-size T-8 model.

its life rubber tyres became an optional extra. The nimble F-12 followed in 1932, initially fitted with a Waukesha engine, as the company did not have a small enough capacity unit. Within months, though, a four-cylinder, 1851cc (113 cubic inch) IHC engine was fitted. The standard tread W-12 made its debut in 1934, along with several other versions such as the I-12 industrial and the Fairway-12. In 1932 the older 15-30 was upgraded as the W-30, and in 1934 the International Harvester W-40 was put on sale – the company's first six-cylinder engine tractor. Just two years later the McCormick-

Deering WD-40 was introduced – the first-ever American tractor to run on diesel.

Early attempts to make a crawler were based around the 10-20 and 15-30 tractors, but in 1931 the company presented the TracTracTor, designated the T-20. Its power unit was the basic Farmall F-20 engine, and it remained in production up to 1939. In the meantime, the International Harvester T-35 six-cylinder petrol model and TD-35 four-cylinder diesel versions were also launched. Further models were seen: T-40 and TD-40, the TD-18 for 1938 and the T-6 and TD-6, launched in 1939.

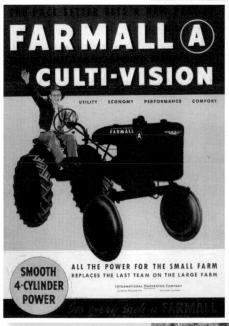

Introduced in 1939, this all-new, streamlined Farmall A used a four-cylinder engine. It was aimed at the smaller jobs on the large farm, and large jobs on the small farm.

International Harvester

The IH plant in Neuss, Germany, was purchased in 1908. In the 1950s the plant expanded to increase production of the DED, DGD4 and DLD2 lines. Shown here is a DGD4.

This is the four-cylinder engine of the DGD4 model, which was a new range of tractors being produced at the International Neuss factory in Germany.

German tractor production started on March 15, 1937, with a Type F-12 rated at 12–15hp. Between then and 1940, 3,973 units of the F-12 family were produced. 1939 saw the public introduction of the second-generation Farmall, known as the signature letter series, with A and B signifying small, H mid-size and M large size. The modern styling of these tractors reflected the efforts of industrial designer Raymond Loewy, who was hired to give the new Farmall tractors and crawlers a distinctive, modern family appearance, with both of these product lines now sharing the same radiator grille design. Loewy also went on to design the prototype dealer facility, with vertical pylon, that unified the look of all 800 US dealers and company stores. Variations of all the letter series tractors were made, and even from its introduction in 1939, the Farmall M became one of the most popular tractors in America.

Wheatley Hall in England was purchased in 1938, but in 1940, with the outbreak of World War II, the factory was requisitioned by the Ministry of Supply to assist in the war effort. Due to allied activities during the war, production at Neuss, Germany, was affected on several occasions, but despite these setbacks, 1945 was the only year in its 60-year history of tractor production that no tractors were produced. Post-war, the tiny Farmall Cub went on sale, using a 983cc (60 cubic inch) four-cylinder engine. The Doncaster site was also returned to IHGB, with implement and service parts production

The Farmall 450 was introduced in 1956 and remained in production through to 1958. It came with a choice of four-cylinder diesel or LP gas engines.

Produced at the IH factory in Doncaster, England, this is the International 634 tractor. It was manufactured between 1968 and 1972.

Tackling the heavy soil is the British-made International B275 model, which uses a four-cylinder engine of 2359cc (144 cubic inches) and puts out around 38hp.

This farmer is happy for his son to be operating this little Farmall Cub. Smallest of the Farmall range, this model was introduced in 1947 and became very popular.

commencing. Tractor production finally started at Doncaster in 1949, when a McCormick International Farmall M, serial no. 1001, was driven off the line on Sept 13. In January 1951 tractor production was started at the new French factory at St Dizier, with the assembly of the Farmall Model FC.

During 1954, IHGB acquired the old Jowett car works at Idle in Bradford, and started production of the McCormick International B-250 tractor. This was rated at 30hp and was Britain's first tractor incorporating disc brakes and a differential locking system. In early 1965 all farm tractor assembly was transferred to the Carr Hill factory, but component parts for McCormick International B-450 and B-614 were still made at Wheatley Hall Road, and transported daily by road. The first tractor off the Carr Hill production line was a McCormick International B-450 in May of that year. In 1968 IHGB launched the McCormick International B-634 tractor – Britain's first model incorporating lower link torsion bar hydraulics, and the last tractor produced in Britain to use McCormick as part of its brand name. 1972 saw the closure of the Liverpool facility at Orrell Park, with work transferred to the remaining British plants, and the last tractor off the N factory line to carry the McCormick name was a 624 in 1972. IHGB started building a new tractor series at their Carr Hill factory in 1970, and in October of that year the first of the 'World Wide' range rolled off the production line. The range initially comprised

The International Harvester B414 Diesel Tractor was launched in 1961. It was powered by a 36hp diesel engine, and transmission was through an eight forward and two reverse speed gearbox.

By the mid-1970s cab design was taking a leading roll, and the International Pro-Ag 86 series of tractors were fitted with what was a state-of-the-art version.

the 454 and 574, followed by the 474, 475 and 674 agricultural tractors, with the 2400 and 2500 industrial machines also included. These had synchromesh transmissions and were Europe's first models offering the option of hydrostatic transmission.

Larger machinery was being produced in the USA, and the early 1970s saw the unveiling of the 66 series, with new rubber-mounted cabs, radios and eight-track sound systems. The race for ever-more power was on, and International used one of their V-8 truck engines to power the 1468 model. Even bigger was the 4366, based on the Steiger machines, 28 per cent of which had been bought in the early 1970s. 1976 saw the Pro-Ag 86 series of two-wheel drive machines introduced, and in 1978 IHGB launched their 84 series tractor range, featuring flat deck cabs and four-wheel-drive. The 85 series, available with the new XL cab or low profile L cab, was launched in 1982 and continued in production until 1987.

In late 1984 Tenneco, owner of the Case and David Brown brands, declared its intention to purchase certain assets of International Harvester's Agricultural Division. The deal was concluded in 1985, and Harvester was placed under the control of the Tenneco Case division. Following this, all products from the Case agricultural division were rebranded Case International, the International Harvester name and its various derivatives gone... or were they..?

The last of the big International four-wheel-drive monsters. This is the Model 7488, which was was often referred to as 'Super Snoopy' due to its long snout.

JCB

The phenomenal global success story of JCB started in Staffordshire, England, from humble beginnings.

The late Joseph Cyril Bamford CBE – known universally as Mr JCB – launched the construction and agricultural equipment manufacturing company that bears his initials in October 23, 1945. He made his first product in a rented lock-up garage that measured 4.5 x 3.6m (15 x 12ft), in Uttoxeter, Staffordshire, and with the help of an inexpensive welding set and some wartime scrap, he produced a screw-tipping trailer, which he sold at the town's market.

By 1948 he had turned his attention to producing a hydraulic machine – Europe's first hydraulic tipping trailer – which was then developed into a hydraulic arm for use on tractors, called a Si-draulic. The first product to carry the JCB initials came in 1953 on a backhoe loader, now known to all as the infamous JCB. The company went from strength to strength, with Mr JCB using his marketing skills to the full. The 1960s saw the 3C presented, which also became a big seller, and publicity stunts were carried out to show how

Mr JCB – Joseph Cyril Bamford – was the man who had a dream and made it come true. Today JCB is a worldwide product that is instantly recognizable.

Bringing one of these back to looking like new is quite a task. This is a JCB 1 gravedigger, and it took two years to restore. It is now fully functioning.

JCB

The compact CX range. The 1CX and 2CX models are versatile and agile, and best suited for the smaller jobs, while the 3CX and 4CX can cope with larger duties.

This is a view under the bonnet of the JCB 1 gravedigger. Its main area of work can be left to the imagination. It is a 1964 model and uses a diesel engine.

manoeuvrable and versatile the machines were. One manoeuvre was to run a car under a machine that had its feet fully extended. Today they not only extend their feet, but can even roll on their sides – although that is not recommended.

In 2005 JCB produced its smallest-ever Mini CX model, which can be easily transported on a trailer. Powered by a 20hp diesel engine, it has a hydrostatic drive on the rear axle. The CX range has a selection of machines, the biggest being the 4CX, a machine for high productivity.

The array of JCB wheeled loaders runs from the 403, a compact and comfortable machine, through to the huge 457 ZX, equipped with the latest 8.9 litre, six-cylinder Cummins QSL engine. The JCB Fastrac is probably the best-known tractor that the company produces. It has a unique suspension system, giving an unrivalled level of comfort and control for extra speed and productivity. These are just a few of the company's products. JCB is privately owned by the Bamford family and the Chairman, Sir Anthony Bamford, is the elder son of the late Mr JCB, taking over the reins in January 1976 – Mr JCB died in 2001, at the grand age of 84.

JCB is the world's third biggest construction equipment manufacturer with 22 plants on four continents, including a Chinese factory in Pudong,

Sand is not the easiest of material to clear, but here a JCB 3C Mk2 Backhoe Loader is seen clearing land for a new development in the Middle East.

With a maximum engine power rating of 216hp, this huge model 456 XZ takes the top spot in its model range.

The Fastrac series of tractors are all-round workers. Seen here with a set of ploughs, Fastrac can cover far more ground than a conventional tractor, in a lot less time.

The latest Fastrac 3230 XTRA offers high power along with the unique suspension system, giving ultimate comfort and control.

JCB has grown from a one-man outfit to a well-respected worldwide company. This is a view of their facility in Brazil.

an Indian factory that opened in 2009 and a £63 million expansion to their Brazil facility that was opened in 2012. JCB exports 75 per cent of its UK-made products to 150 countries. The company's worldwide headquarters are based in Rocester, Staffordshire.

The company manufactures 300 different machines in 15 product ranges. For agricultural markets, they produce a range of telescopic handlers and the unique Fastrac tractor, along with the Teletruk forklift.

In 2003 JCB announced that it intended to develop and manufacture its own diesel engines, and JCB Power Systems Ltd was formed to produce the engines. In 2006 the JCB Dieselmax broke the world diesel-powered land speed record using two specially tuned JCB444 diesel engines.

From humble beginnings, JCB is now a truly international company that continues to expand and innovate as it approaches its 70th anniversary.

The bare bones – this Fastrac model has no distractions. It is versatile enough to have all kinds of implements attached to it, both front and back.

John Deere

The history of the John Deere tractor company goes back to a period when early settlers were moving to the Midwest of America, where there were vast open plains just waiting to be farmed and cultivated.

John Deere (1804 –1886), the founder of Deere & Company. Today it is one of the largest agricultural and construction equipment manufacturers in the world.

It was the Waterloo Boy tractor that gave the John Deere company its entry into the tractor market. It used a two-cylinder engine that burned kerosene.

Back in 1893, John Froelich – credited with being the creator of the first gasoline tractor – and a group from Waterloo, Iowa, formed the Waterloo Gasoline Traction Company, to market the Froelich machine. Four experimental tractors were built, of which two were sold and then returned. In 1895 the company reorganized itself as the Waterloo Gasoline Engine Company. In 1902 they and the Davis Engine Company formed the Waterloo Tractor Works; their aim was to manufacture gasoline engines and three-cylinder automobiles. Withdrawing from the company just two years later, the Waterloo Gasoline Engine Company presented a redesigned engine called the Waterloo Boy. These were produced in 1913, starting with the Model LA. The Model R followed, and by 1915 more than

John Deere listened to the pioneer farmers, and took note of the problems they encountered with the fertile soil they were trying to work with.

Deere's Walking Cultivator.
Patented August, 1867.

The John Deere walking cultivator was patented in August 1867. Although many farmers preferred to ride, this walking unit was far less expensive.

FACT BOX

John Deere, the pioneer

- **1804** February 7 – John Deere is born in Rutland, Vermont, USA.
He serves a four-year apprenticeship learning to be a blacksmith.

- **1825** John Deere starts his career as a journeyman blacksmith – his highly polished hay forks and shovels become much in demand.

- **1836** John departs Vermont and arrives in Grand Detour, Illinois, to join fellow settlers from Vermont, leaving his wife and family behind, to join him later.
A forge is built and work starts – shoeing horses and oxen, repairing ploughs and other equipment.
John learns of the difficulties that the pioneer farmers have in cultivating the fertile land of the Midwest – the heavy soil clogs the blades of the plough and precious time is wasted, constantly having to clean them off.

- **1937** John sets to work on a highly polished and properly shaped moldboard, which would clean itself as it turned the furrow slice. This is tested on the Lewis Crandall farm nearby, and is a great success. Two further models follow, and the John Deere name and his 'self-polishers' spreads throughout the West and beyond.

- **1841** First emigrant train of covered wagons reaches California.

- **1843** John starts importing steel from the mills in Sheffield, England.

- **1846** Jones and Quigg Steel Works of Pittsburgh produces the first slab of cast plough steel made in the United States.

- **1848** John moves to Moline, starts a partnership with John Gould and Robert Tate, and constructs a new building.

- **1849** The workforce total is about 70, and they make 2,136 ploughs this year.

- **1850** Company is now called Deere, Tate & Gould.

- **1852/68** John buys his partners out. The company changes its name several times: John Deere; John Deere and Company; Deere & Company; and the Moline Plow Manufactory.

- **1853** John Deere's son Charles joins the company as bookkeeper at the age of 16.

- **1858** John Deere hands over the reins of leadership to his 21-year-old son Charles.

- **1861** American Civil War starts. Large-scale Midwest farming commences, to keep armies and animals fed. This also results in the improvement of farm machinery.

- **1863** Deere produces the Hawkeye riding cultivator, under license from the inventor Robert Furnas.

- **1865** Demarius Lamb Deere, John's wife, dies at 60 years old.

- **1866** John marries Lucenia Lamb, sister of Demarius.

- **1867** The Walking Cultivator is patented.

- **1870** Deere has five basic lines to see them through to the end of the 19th century: ploughs, cultivators, harrows, drills and planters.

- **1875** The Gilpin Sulky (riding) Plow is developed – probably the most significant piece of machinery to be presented by Deere in the post-Civil War years.

- **1876** Introduction of the 'leaping deer' trademark.

- **1886** John Deere dies at the age of 82. His son Charles takes over the presidency.

- **1907** Charles Deere dies.

- **1916** Deere start building experimental/prototype tractors, but these never go into production.

one hundred were sold. The Model N was introduced in December 1916, and included enclosed transmission and two forward speeds. Both the R and N ran on kerosene, and had two-cylinder engines. By 1918 they had a tractor manufacturing facility, filled with machines, in Waterloo, Iowa.

The Deere & Company board had shown a reluctance to making tractors, but on March 14, 1918, by a unanimous resolution, they agreed to purchase the Waterloo Gasoline Engine Company. With these events, the Deere Company started selling its first commercial tractor.

The following years were riddled with problems. A strike and then falling sales of the Waterloo Boy tractors resulted in lay-offs, wage restrictions and product price cutting. The Model N had had little in the way of updates during its life, and the time had come to remedy that. As Model N serial number 30400 rolled off the production line in

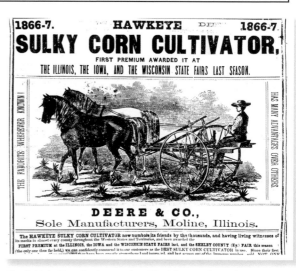

The first John Deere implement adapted for riding was the Hawkeye Riding Cultivator, made in 1863. This is a similar but later model.

John Deere

The Gilpin Sulky (riding) Plow was developed and patented in 1875. It was an instant success, and by the late 1870s it was one of the largest-selling sulky ploughs in the country.

The Froelich tractor, which was also the forerunner of the later Waterloo Boy tractor, is seen by many to be the first successful petrol tractor.

This is the first John Deere trademark, depicting the leaping deer. It was registered in 1876, even though the registration papers show that the mark was used earlier.

June of 1923, it was followed not by another Model N, but by the new Model D, with serial number 30401. This was the first production tractor with the John Deere name on it and used an improved two-cylinder, horizontal, water-cooled Model N engine. The tractor was a great success for the company and remained in production until March 1954. By the late 1940s, electric starting and lights had become an option. Farmers loved the simplicity and low cost maintenance of the machine, and the fact that it would run on most kinds of cheap fuel, while still being capable of pulling a three-bottom plough in most conditions. Some of the last examples of the Model D were assembled on the roadside next to the factory, and for this reason they became known as 'streeters'.

The design for the GP (General Purpose) model was started in 1925, and in 1929 the GP Wide-Tread row-crop model was put on the market. It was fitted with an exclusive mechanical lift that was foot-operated – a John Deere first. Although started on petrol, it could then run on paraffin; an XOGP model was produced for 1930 – XO denoting a new crossover-style manifold.

Management of the Deere company was transferred to John's son Charles, who had a shrewd business sense. His father remained president of the company until his death in 1886.

Waterloo Boy tractors were exported to England and Ireland, but took on another name. Although the same machine, they were known as the Overtime.

In 1918 the Deere company purchased the Waterloo Gasoline Traction Engine Company of Waterloo, Iowa. This is the pretty company emblem.

The John Deere Model D became the longest-running model ever produced. It was presented in 1923 and was still in production up to 1953.

A rare 1917 period scene showing a farmer and his Waterloo Boy Model R, hard at work ploughing the field, ready for the crops to be planted.

The economic downturn had by now worsened, and the Great Depression started to hit hard. Once again, wages were cut, there were more lay-offs, pensions were cut, and working hours were shortened. By 1933, business was poor and sales were plummeting, but somehow the company struggled on. Despite all the doom and gloom, the new Model A tractor was produced for 1934, followed a year later by the smaller version Model B.

One new feature of the A was the adjustability of the rear wheel track. With rear wheels generally fixed at 102–107cm (40–42in) apart, they could now be adjusted by sliding them in and out on their splined axles. The styling on the Model A was changed for 1939, and electric starting was added as an option, only to be made a regular feature by 1947. The smaller Model B was advertised as a one-bottom plough tractor giving the daily work output of six to eight horses, and aimed at either the larger farmer who needed a tractor to carry out smaller tasks or introducing the smaller horse-bound farmer to the idea of a small and

John Deere

Although this has the John Deere name and colours, it is in fact a conversion made by Lindermann Brothers, mounted on the BO tractor chassis.

A view taken during 1936, where a job-lot of 45 John Deere tractors are for sale. Some even have the names of the buyers on them.

inexpensive tractor. This model sold so well that it became the best-selling two-cylinder John Deere tractor ever produced. In 1935 John Deere paired up with the Caterpillar Company, who had a very strong crawler division. The two companies decided to sell each other's products, with particular emphasis in California. As the Depression started to lift and the 1930s came to a close, industrial designer Henry Dreyfuss started work with the engineers at Deere to streamline the A series tractors. From now on, more attention would be paid to the aesthetics of a John Deere design, as well as its utilitarian requirements.

In 1923 the Model GP was introduced, with arched front axle for high crop clearance. For 1932 a new wide-tread version was also presented, as seen here.

The John Deere Model AR, manufactured at their Waterloo facility and produced between 1935 and 1952, could be run on either kerosene or petrol.

Made between 1934 and 1952, this is the John Deere Model A general-purpose, two-plough tractor. It was the first to have an adjustable wheel tread.

would eventually turn into the Model R, presented in 1949. This was the company's first diesel tractor and the first to have live PTO and a cab.

Designation letters were now replaced by numbers: the Model 40 replaced the M, the Model 50 replaced the B, the Model 60 replaced the A, the Model 70 replaced the G, and the Model 80 replaced the R. The Models 50 and 60 were introduced in June 1952, and the 70 came in April of 1953, using a gasoline or all-fuel engine. For 1955 the Model 80 was introduced, catering to farmers' requirements for more horsepower. This was a stop-gap machine until a new, more powerful and larger tractor could make its debut in 1960; the era of the two-cylinder engine was coming to a close. More powerful than both the Model R which it replaced, and the 70 diesel, the 80 had a water-cooled, twin-cylinder, 7783cc

The last batch of 92 Model D tractors were assembled on the road next to the factory. In this way, they acquired the nickname 'streeters'.

Late in 1937 the Model L replaced the older Model 62, still with a twin-cylinder engine, although now it was vertical rather than horizontal; it also featured a three-speed transmission. 1941 saw the larger Model LA, which had more power, weighed more, had higher crop clearance, and optional electric starting and lights.

During the war years, Deere produced military tractors, ammunition and aircraft parts, along with other items for the war effort. Some 4,500 employees went to war, many of whom served in the 'John Deere' battalion – a specialist ordinance group who also saw action in Europe.

In 1945, with the war now over, 295 hectares (730 acres) of land was acquired near Dubuque, Iowa, and the building of a new factory was started. During 1942 and 1943, work had started on the Model 69, which turned into the Model M, presented in 1947. Two years later came the MC, a crawler version of the Model M, with Touch-o-Matic hydraulics available as an optional extra on the MT model. A new Model MX was developed over a period of several years, and it was this that

John Deere

The John Deere Model 70 was produced between 1953 and 1956, and could use petrol, tractor fuel, LPG or diesel to power its engine. Seen here is the LPG tank.

The Model L tractor, seen here pulling a John Deere trailer, was introduced in 1937. It was classed as a lightweight, economical, two-cylinder machine.

January 12, 1949 finally saw the introduction of the first Model R diesel tractor. Some 17,000 of these fuel-efficient machines were sold during their lifespan.

(475 cubic inch) engine, and six-speed transmission. During this time, the Model 20 was also being developed, but, more importantly, so was the New Generation of Power tractors, which would make their debut in a blaze of glory in 1960.

In 1955 William A. Hewitt was elected president and later became CEO; he would be the last of the Deere family to head up the company. The following year John Deere built a factory in Mexico, to produce a small tractor. A presence was also made in Spain, and a harvesting company was bought in Germany. In 1956 John Deere and Lanz became international partners, with John Deere's acquisition of a majority shareholding, and in the next few years Deere also moved into France, Argentina and South Africa, thus making them a truly international company.

The early numbered series were replaced in 1957, with the Model 40 becoming the 420,

1968 saw a more streamlined version of the leaping deer logo. The silhouette shows just two legs rather than the four, and only one set of antlers.

and so on. A completely new Model 320 was introduced, based on the M and MI models, built at the Dubuque factory. In 1958 the 20 series was replaced by the 30 series – the 420 now became the 430, and so on. A further new series was released in the form of the 435 for 1959, when the company also introduced the 8301 and 8401 models, aimed exclusively at the industrial market and fitted with a Hancock scraper system. At the end of the 1950s the company introduced the 8010, a diesel-powered 215hp tractor, much ahead of its time. This was their first four-wheel-drive, but unfortunately only few were sold, all of which had to be recalled, updated and returned as 8020s.

As 1960 dawned, a new 10 series was created, with designations starting with the 1010 and 2010 models. These were not new models, but merely the two-cylinder tractors with new engines, although Models 3010 and 4010 were entirely new machines and became known as the 'New Generation of Power' tractors. These machines piloted John Deere into prime sales position, overtaking their biggest rival, International, for the first time. The 3010 used a four-cylinder engine, while the 4010 had a six-cylinder unit.

Although the little Model L continued to use a two-cylinder engine, it was a vertical configuration unit rather than the traditional horizontal.

In 1942 the Model G was 'styled', and then became the Model GM. It was introduced in 1938 as the biggest row-crop tractor produced by the company.

Seen here are employees of the John Deere company, who enlisted in the John Deere Battalion. Some 4,500 people left the factory to serve their country.

John Deere

The Model 530, like its smaller 520 stable mate, was only made in general-purpose guise. There was also a choice of front-end: tricycle, single and wide.

These were diesel engines, although both petrol and LPG engines were also available. With the addition of power steering, power brakes, power implement raising capabilities and an eight-speed transmission, these machines became extremely popular. The following year, a 5010 model – the industry's first two-wheel-drive tractor rated over 100hp – was introduced with 151hp. Three years after their introduction, the 3010 and 4010 were replaced with an updated 3020 and 4020 – the latter becoming the most popular tractor of its time.

Construction started on a new engine factory in 1961 at Saran, near Orleans in France, while at the same time the Deere & Company administration centre was also being built in Moline. The following year, the company celebrated its 125th anniversary, when they also bought a majority interest in South African Cultivators, a farm implement company based near Johannesburg.

For 1972, the same year that the 'Green Girl' was born, the 'Generation II' tractor model line was launched. New styling, Perma-Clutch and a Sound-Gard cab all

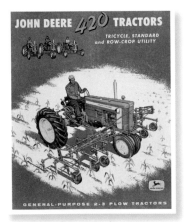

The advertisement for the John Deere 420 tractor tells it all – a general-purpose and very versatile machine.

The John Deere Model 4010, which was introduced in 1960 and lasted through to 1963, had a choice of petrol, diesel or LP gas engines.

Produced between 1958 and 1961, this is the John Deere 830 diesel. Produced at the factory in Waterloo, Iowa, the engine had a displacement of 7713cc (470.7 cubic inches). Some 6,892 examples were manufactured.

Seen here is a Model 4010 diesel tractor going about its harvesting duties, with a John Deere combine attached to the back.

went to make these an instant success for the company. Always at the forefront when it came to safety, John Deere was now putting emphasis on cab comfort and safety. These cabs were comfortable, sealed from the dust and dirt, and air-conditioning or heaters were optional extras. Tinted glass to stop glare, along with seat belt and adjustable steering wheel, went to make this cab extremely desirable.

The new 30 series tractors consisted of the 80hp 4030, the 100hp 4230, the 125hp 4330 and the 150hp 4630. Huge crop failures outside the United States spurred massive foreign buying of American grain, and with it farmers prospered dramatically. The demand for equipment also exploded, and John Deere total sales increased to a staggering

John Deere

$2 billion for the first time. This same year – 1973 – the board at John Deere decided on a more independent board structure, and the first outside director was appointed.

In 1977 an agreement with the Japanese manufacturer Yanmar allowed the sale of small tractors under the John Deere name, along with a new Canadian headquarters in Grimsby, Ontario, a new engine works in Waterloo and new sales branch offices in Atlanta; company employment reached an all-time record high. Sales topped the $5 billion mark with earnings hitting a record $310 million – what a way to end the 1970s.

The 1980s kicked off with the introduction of a four-row cotton-picker, and the following year two new super tractors were unveiled – the 40 series, four-wheel-drive machines now being replaced by the 50 series 185hp 8450 and the 235hp 8650 six-cylinder models. Farmers with large holdings were now looking for bigger and more powerful machines to do their work. In addition to these, John Deere added the even bigger 8850, a

A scene taken in 1962, this is a John Deere Model 3020 with a rear-mounted cultivator.

This operator is ploughing the field in an opposite direction to the previous plough lines, to help turn the soil. He is riding a John Deere Model 4020 diesel.

new power-house model producing 370hp. This tractor used a turbocharged and intercooled V8 diesel engine that was designed and built by John Deere themselves at the Waterloo plant.

Following the bumper 1970s came a 1980s that would see a deepening recession, and the advertising for these machines reflected this when they were described as 'three new ways to tighten your belt'. You could be forgiven for thinking there was no recession going on, though, as this was the most expensive John Deere model, and it was also the biggest to date. Fortunately, all the investment in the new plant, facilities and machinery that the company had instigated in the 1970s paid off during these recession years. Updated versions of the Mannheim-built tractors were also presented, ranging from the three-cylinder 2150 model through to the larger six-cylinder 2950 model. The Yanmar connection also came into play when Deere started importing one of these machines as the 1250 model, in corporate colours. Bigger models followed, and some specialist models were also presented. Other new machines followed, and older models were updated, redesignated and improved as the 1980s and the recession moved to a close.

In 1993 the new 5000, 6000 and 7000 series tractors were presented, helping to put John

This is the Model 7520 super tractor, produced in 1970. This was an uprated version of the 7020, and was fitted with four-wheel drive.

The Model 4250 was made between 1983 and 1988. It used a 7636cc (466 cubic inch), six-cylinder, turbo-diesel engine.

The 7520 was aimed at the large acreage farms. Its 8701cc (531 cubic inch) capacity engine provided the additional power customers wanted.

John Deere

The Model 8400T was manufactured at the John Deere Waterloo facility. The 'T' in the designation denotes a tracked vehicle.

A huge tractor in which its operator is dwarfed – this is the powerful Model 6810 of 1993.

Deere at the head of the table for tractor sales in the German market. The 5000 series used a three-cylinder engine, had hydrostatic power steering, nine-speed transmission and optional four-wheel drive. The four-cylinder 6000 series tractors were produced in the Mannheim factory, Germany, while the six-cylinder 7000 series were built in the USA at a new factory in Atlanta. There was a state-of-the-art cab to fit all models, claimed to be the quietest on the market, with more space and more glass area.

For 2005 the line-up of tractors being produced by John Deere was sufficient to cater for pretty well any need, ranging from their compact 23hp 2210 utility tractor range, which John Deere class as their most compact tractor ever, through to the 8020 series tractors. These tractors lead the way in power with up to 255-PTO horsepower, exclusive ActiveSeat and front independent link suspension providing outstanding operator comfort. Five-wheeled and five-tracked models are included, the tracked models having a 'T' after their designation.

This style of tractor, or telehandler, as it is known, became popular during the 1990s. This is the Model 3400, made for John Deere by Matbro.

Advertising encouraging you to fit the extra wheels – the more you spread that weight, the less compaction you have. This is the Model 9300.

The 7810, made between 1997 and 2003. It has a six-cylinder, turbo diesel engine and PowerQuad transmission, with 16 forward and 16 reverse gears.

Introduced in 2011, the 8360R uses the new model designation system with the last three digits showing the engine horsepower size, so it is 360hp for this 8000 series machine.

Taking tractor productivity and performance to the highest levels in 2004, John Deere introduced the 9620 and 9620T tractors rated at 500hp. These four-wheel-drive and large-track tractors were the largest ever built by John Deere and were powered by the 12.5 litre PowerTech engines. This was surpassed in 2012 by the 9560 with a 13.5 litre, 560hp engine.

There is no reason to question why John Deere is the biggest farm machinery company in the world. Their products and equipment are extensive, their tractors are probably the best in the world, and the company heavily invests in research and development. They have outlets all over the world, and whether you are looking for a mower to trim your lawn or a huge monster of a tractor to use on your farm, they can cater for these and pretty well anything in between. The company has come through recessions, farm worker unrest and stock market crashes. It is still with us today and will no doubt still be with us for a long time in the future – these roots go back a very long way.

Lamborghini

The Lamborghini company, named after its creator Ferruccio Lamborghini, is better known for its exotic sports cars. However, the company did not start by building high-performance cars; in fact, it was building successful tractors long before that.

This early Lamborghini model is the type that would have served the small Italian farmer so diligently, and which are still seen.

Lamborghini was born into a farming family, and when he returned from World War II, he dabbled with converting old Morris military engines and fitting them into tractors, which were much needed at that time. This work was initially carried out in an old barn, but business was brisk, and by 1949 he had moved into larger premises. When he had a new factory built, it saw the birth of Lamborghini Trattori SpA in Cento, not far from Bologna, Italy. The Morris engines were soon being replaced by MWM, Perkins and even Lamborghini's own units, and it wasn't long before the L33 model tractor was introduced. Starting off building one tractor per day, the company grew over the next few years, and by 1958 they were producing 1,500 tractors per year. Eighty per cent of the tractor parts were now produced at the

The huge Model 115 Formula tractor produced by Lamborghini used a six-cylinder engine and produced 115hp. With 36 forward and 36 reverse gears, it coped with most duties it was required to carry out.

The Lamborghini Model R 603 DT was manufactured in the early 1980s, and was a four-wheel-drive machine.

This Crono 70A is a middle-of-the-range model which was made between 1993 and 2001. It uses a three-cylinder, turbo-diesel engine.

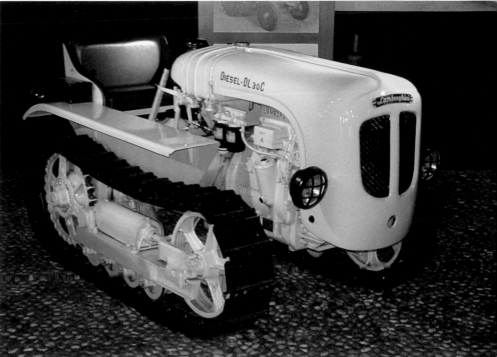

Lamborghini not only made tractors, but they also made crawlers such as this Model Cingolato DL 30C, which is diesel-powered.

factory, allowing Lamborghini to keep an eye on the quality of the materials. The tractors gained a good reputation, being reliable, well made and very robust. They were regularly seen at tractor pulling events that Lamborghini himself organized for the local farmers.

By 1969 the company was producing some 5,000 units per year, and a further move was needed to cope with the increase in demand. This move took place in 1971, by which time Lamborghini was the third-best-selling tractor in Italy.

A crushing blow to the company came in 1972, which led Lamborghini to rethink his commitment to the tractor business. A large order of tractors from a Bolivian company fell though, and this event became one of the critical factors in Lamborghini deciding to sell his business, which he did to the SAME group of Trevigliano. Continuing to thrive under its new owners, however, 1979 saw the company producing 10,000 tractors per year, worldwide.

The current Lamborghini tractor range comprises the R1 models, the smallest machines, which have excellent engine performance and outstanding manoeuvrability. At the other end of the scale, the company announced a new line-up in 2013, including the new Mach VRT, the flagship machine with up to 265hp and a variable transmission, and the Spark series ranging from 120–190hp. These join the 90–130hp Nitro range powered by the Deutz Tier 4i engine to comply with the latest emmission regulations.

Seen here is the 141hp Model R6, which uses a Deutz 2012 engine, and has 40 forward and 40 reverse gears.

Landini

Landini is the oldest established tractor company in Italy, and was founded by Giovanni Landini in 1884. The company has seen turbulent times, but today produces an exemplary range of machines.

A close-up view of the engine of the Model L25. This is a Testa Calda one-cylinder unit of the 1950s.

As a blacksmith, Giovanni Landini set up his workshop in the small town of Fabbrico, near the Po river in the Reggio Emilia province of Italy. His skills soon moved him into manufacturing agricultural implements and wine-making equipment, shortly to be followed by steam engines and internal combustion engines.

In 1910 Landini produced a 'testa calda' type engine, or hot bulb engine, often referred to as a semi-diesel engine.

Landini was never to see the fruits of his work though, as he sadly passed away in 1924. All was not lost, however, as his sons took over the running of the business, and from here the foundations that their father had laid would grow to be the successful Landini company of today.

It was only a year after their father's death that the first Model 25/30 prototype was presented, which became the first authentic Italian tractor. The Landini 25/30 of 1925 was a well-made

The 2000 year model Landini Legend 185. These machines use a Perkins, green diesel engine that offers exceptional reliability and performance.

It was 1924 when the Landini company introduced its Testa Calda (Hot Bulb) engine. Seen here is the Model L25 of the 1950s.

108

Landini

Built in 1985, this is the Landini 10000S model. It is a powerful tractor with four-wheel-drive, and is able to produce 105hp.

With its Perkins AT 4.236 model diesel engine, this is the Landini Model 9880, which was manufactured in 1988.

machine and was not only easy to use but, more importantly, repairs could be carried out easily by the farmers themselves. The engine was a single-cylinder, two-stroke, hot bulb model, which was water-cooled. It was equipped with four forward gears and one reverse gear, which made it very versatile and able to cope with many different types of terrain, and a variety of workloads.

This tractor was upgraded to 40hp in 1928, and in 1934 the SL 50 Super Landini model was announced. This produced 48hp and was the first tractor to use a radiator. Again, it had a single-cylinder, hot bulb engine with just three forward gears and one reverse.

This was followed in 1935 by the VL 30 Velite models, which were produced through to World War II. Post-war, production was restarted, and the L25, a 4300cc (262 cubic inch) engine, 25hp model was announced. This machine was built at a new factory in Como.

For 1955 Landini produced their most powerful hot bulb model tractor, the 55L, but the hot bulb engine was on borrowed time, and two years later, after an agreement with Perkins engines, Landini gained a license to produce and fit their engines.

This agreement has continued to the present day, when Perkins engines are still fitted to the current range of machines.

In 1959 Landini moved into the crawler market with the introduction of the C35 model. It was also this year that the company was taken over by Massey Ferguson, who were keen to exploit the crawler division of Landini. The company did well under its new owners and a new plant was opened in 1968 at Aprilla, to manufacture heavy-duty industrial construction machinery.

New 6500, 7500 and 8500 models were introduced in 1973, with redesigned transmissions, giving the tractors 12 forward and four reverse gears. This was followed in 1977 by the first 100hp Landini model, a four-wheel-drive machine that used a Perkins straight six-cylinder engine, and the range was increased to include outputs from 45hp to 145hp.

This is an English-language version of an advertisement for the Landini Blizzard 95, which was manufactured in the late 1990s.

Landini

The 2005 Landini Ghibli 90 was fitted with a new cab, giving better all-round vision, and it was also equipped with air-conditioning.

This crawler is from the Trekker model series. All of these use a fluid-cooled, four-cylinder Perkins diesel engine.

During the 1980s, Landini moved into yet another sector – 1982 saw them introduce their first Orchard model, and in 1986 a new Vineyard model was presented too. This was a whole new area for Landini, but it was not long before they had a large share of that market, becoming sole suppliers of orchard and vineyard versions for both crawler and wheeled tractors, but branded as Massey Fergusons.

With sales of over 13,000 tractors, not only was 1988 a bumper year for Landini, but they also updated their mid-range models, now classified as the 60, 70 and 80 models.

Massey Ferguson made the decision to sell off 66 per cent of its Landini shares in 1989 to the Eurobelge/Unione Manifatture holding company, who in turn later sold their controlling interest in its affairs to the Cameli Gerolimich Group, who finally took Landini SpA into the Unione Manifatture. Now at the forefront of the tractor industry, Landini began redesigning their vehicles, and produced a new series to be announced as the Trekker, Blizzard and Advantage models.

The Powerfarm series models – 75, 85, 95, 105 – feature 68 to 91hp, Perkins 1100, four-cylinder, naturally aspirated and turbocharged engines.

Landini

The Landini Vision series is a range of medium-power tractors that use the Perkins 1100 series Tier 2 engines.

In February 1994 Valerio and Pierangelo Morra, representatives of the ARGO SpA family holding company, became president and vice president of Landini SpA, respectively. Together with Massey Ferguson, they contributed to a substantial recapitalization of the Landini concern. March 1994 saw Iseki join Landini SpA, and an increase in tractor sales of more than 30 per cent over the previous year was announced.

It was in 1995 that Landini acquired Valpadana SpA, a prestigious trademark in the Italian agricultural machinery sector. Overseas markets were starting to open up too, and distribution

The Vision 95 has a two-speed, hydraulically operated PTO, and the powerful hydraulics with up to five auxiliary valves are additional features.

started through the AGCO network in North America. Landini Sud America was opened in Valencia, Venezuela, with the aim of promoting the brand in Latin America.

During 1996 the Fabbrico factory was given an overhaul to cope with the extra production demand. A new assembly line was fitted which allowed double the previous production levels. This same year the new San Martino factory was also opened, concentrating purely on machining parts, such as gears and prototype components.

The following year the Legend II series was launched, along with the Globus range. World distribution continued with importers being opened in Spain and Germany. The following year saw the introduction of the Discovery and Minstral range, with a new transmission system, the Deltasix, being launched in 1999.

As the new millennium came in, new models appeared in the form of Rex Orchard and Vineyard, Mythos, Ghibli, Atlas and new Trekker. For 2013 the Landini range included various models, with power outputs ranging from the 21–55hp 1 series compacts through the 5 series of 90–115hp machines using the latest Tier 4 Perkins 854 engine, and up to the 7 series that combines economy and power through the new SCR (Selective Catalyst Reduction) engine system.

Massey Ferguson

The Massey Ferguson company was created in 1953 through the merger of three agricultural companies: Massey, Harris and Ferguson. The company was later bought out by the AGCO Corporation.

Daniel Massey (1798–1856) was a blacksmith in Newcastle, Ontario, Canada. He began production of agricultural implements in 1847.

The sales success of the open-end binder in 1890 prompted Hart Massey to propose a merger of the two companies. Alanson Harris agreed, and so it was that on May 6, 1891 that the two companies became one, under the name of Massey-Harris Company Ltd. Smaller agricultural companies were also added over the next few years, increasing their versatility and product range. Even so, the company was slow to move into tractor manufacturing.

Although Massey-Harris was assembling Olds engines, they bought the Deyo-Macey engine plant at Binghamton, New York, in 1910, and by 1914 they were making their own.

When World War I started in 1914, farm mechanization was required to step up a gear, and

This is a view of what the Massey foundry looked like. It has so many interesting elements to it, and gives a real feel for the period.

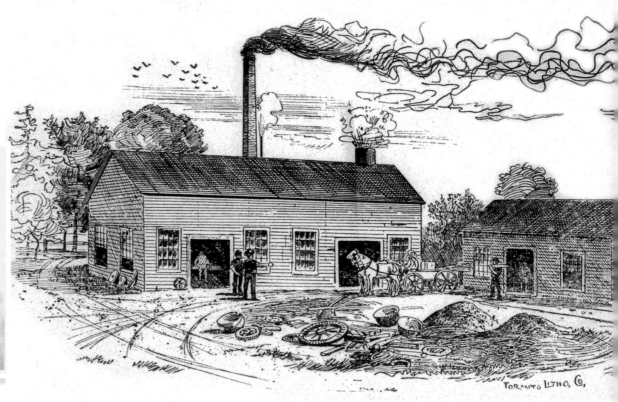

An early Massey Manufacturing Company advertising brochure cover. Clearly noted in the bottom left-hand corner is the fact that they were the oldest and largest in Canada.

LA MÁQUINA DE SEGAR "MASSEY."

This is a very early catalogue from 1881. A rough translation from the Spanish is "The cutting machine from Massey".

An 1891 catalogue depicting the No. 6B Massey-Toronto light binder in action. This was apparently introduced the season before.

THE WORLD'S FAMOUS MASSEY-TORONTO BINDERS

NO. 6B

Massey-Toronto Light Binder No. 6B.

The Toronto Mower

This Massey Illustrated journal of May 1890, produced by the company, depicts the Toronto Mower drawn by two horses.

the tractor became instrumental in this work. Tractors could cover more work in a day than a man and his horse could cover in a week, and with Canada also having troops fighting in Europe, Massey-Harris had to be involved. Unfortunately they had no tractor in production, and so in 1917 a search was started to find one that could be imported into the Canadian market. They chose the Big Bull model of the Bull Tractor Company, who were no strangers to the tractor market.

Sadly, due to a shortage of parts the import agreement between Massey-Harris and the Bull Tractor Company came to a premature end. A new agreement was settled with the Parrett Tractor Company of Chicago in 1918, in which Massey-Harris would build and market the tractor under their own name for Canada and some export markets. Dent Parrett drew up designs for three models, the MH1, MH2 and MH3, and production started in 1919 at the Massey-Harris engine factory

in Weston, Toronto. The models came in three different sizes: 12–25hp, 12–22 hp and 15–28hp. All three used a Buda water-cooled, four-cylinder engine, capable of running on petrol or paraffin. They were fitted across a steel frame chassis and had two forward gears and one reverse gear. Despite being well-made machines, they dated quickly and were made redundant by the new lighter models coming on to the market.

Having now burnt their fingers twice, the company opted for caution and decided to wait to see what would happen within the industry. When people returned from the war, there was little or no work, as a strong recession created havoc around the world. The dilemma Massey-Harris had was that the tractor business was still moving forward at a rapid pace, and all the time they did not have a tractor division, they were potentially losing money.

As the world economy started to recover in the mid-1920s, Massey-Harris started looking for another partner. This time it was the J. I. Case Plow Works Company of Racine, Wisconsin, in 1926, which led to the acquisition of the company the following year. Although Massey-Harris had

---

FACT BOX

Massey company background

- **1630** Massey family emigrate from England to the United States.
- **1795** Parts of the family move to Watertown, New York State.
- **1798** Birth of Daniel Massey Jr.
- **1802** Daniel Massey, his wife Rebecca Kelley and son Daniel move to Haldimand Township near Grafton, Ontario, Canada.
- **1804** Six-year-old Daniel Massey returns to Watertown to live with his grandparents.
- **1819** Daniel Massey returns to Canada and purchases 80 hectares (200 acres) of land near the family farm.
  Massey marries childhood sweetheart Lucinda Bradley.
- **1830s** Massey returns to farming and purchases machinery and tools, not available in Canada, during his trips to the USA.
  The Bull Thresher is bought and set up in a barn – this is also used by neighbouring farmers. A small machine shop is built. Massey turns his farm over to his son Hart.

- **1847** Soon after closing his foundry, Richard Vaughan strikes up a partnership with Massey to build implements in his foundry with Massey money.
  Massey buys Vaughan out.
- **1849** Massey moves to Newcastle and names the company Newcastle Foundry and Machine Manufactory CW.
- **1851** Hart Massey joins his father in the business.
  Hart obtains Canadian patent rights for the Ketchum Mower – the first grass-cutting machine.
- **1855** The advanced Manny Combined Hand-Rake Reaper manufactured.
- **1856** The Grand Trunk Railroad reaches Newcastle, opening up new possibilities for Hart Massey to exploit.
  Products now include farm machinery, steam engines, boilers, castings, stoves and lathes, to name a few.
  Daniel Massey dies at the age of 58.

- **1862** Company is renamed The Newcastle Agricultural Works.
- **1863** March 29 – the warehouse is demolished by fire.
- **1870** The Massey Manufacturing Company is formed with Hart Massey as president and his oldest son Charles as vice president and superintendent.
- **1871** Hart Massey retires.
- **1879** The company moves to Toronto due to lack of facilities in Newcastle.
- **1891** May 6 – Massey Manufacturing joins with A. Harris and Son Company. The Massey-Harris Company is formed.

113

Massey Ferguson

Looking a little like a clockwork machine with grass-cutting attachment, this is the very successful Kirby two-wheel mower.

A. Thompson Rei & Cia of Santiago were apparently the sole agents for the Harris Company products. The Harris name had already spread.

An engraving of the small foundry that Alanson Harris bought in Beamsville, Ontario, Canada, back in 1857.

bought the Case company along with its name, it sold back the rights to the Case name. What they had acquired was the Wallis family of tractors, which were well known both for their excellent fuel efficiency and their distinctive U-frame construction. Further help also came when Fordson decide to move away from tractor production.

Massey-Harris continued the production of the Wallis tractors and established the old Case factory in Racine as their own, and were now in a position for the first time to infiltrate the American market. The current Wallis tractor was the 20-30, which used a 5669cc (346 cubic inch) water-cooled, four-cylinder engine with a two-speed transmission. A smaller version of essentially the same machine was produced, the 12-20, and the Model 25 followed in 1931, which was also known as the Massey-Harris 26-41.

Work also started on the first true Massey-Harris, the General Purpose machine, which unfortunately did not live up to its name. Besides the coming of the Great Depression, which would devastate the farming community like never

Alanson Harris (1816–1894) established a foundry at Beamsville, Canada, in 1857, and then relocated to Brantford in 1872.

The Massey-Harris Number 2 was a slightly improved version of the Number 1 model. It ran on petrol or kerosene, and had enclosed rear-wheel gears.

THE No. 9 OSBORNE SELF-BINDING HARVESTER

A depiction of the Osborne self-binding harvester, taken from a catalogue of 1880. It is difficult to understand what these machines meant to the farmer.

FACT BOX

Harris company background

- **1816** Alanson Harris is born.
- **1857** Alanson Harris, now a farmer and mill owner, founds his implement business at Beamsville, Canada.
- **1863** Alanson is joined in the business by his son John. Kirby mower and other US agricultural machinery are acquired. A. Harris and Son Co. becomes a big competitor to the Massey Company.
- **1872** A. Harris and Son Co. moves to Brantford.
- **1880s** The Brantford binder becomes one of the company's best-selling lines.
- **1890** A. Harris and Son Co. introduces the open-end binder – a modification of the D. M. Osborne company design.

As mentioned on the illustration, this is a birds-eye view of the A. Harris, Son & Co. works in Brantford, Canada.

Two new machines came in 1936, the Challenger and the Pacemaker, which used the same engine and similar components. In 1937 the Twin Power Pacemaker was introduced, having a more streamlined and enclosed engine cover. 1938 saw a general update to the range and the launch of the Model 101, which had a long, sleek bonnet under which was incorporated a new Chrysler 3293cc (201 cubic inch), six-cylinder truck engine. A cast iron chassis now substituted the old U-frame, and the following year it was joined by the 101 Junior, a smaller version. Other derivatives followed, such as the 201, 202 and 203 models. The 81 and 82 followed in early 1940, both smaller, lighter and less expensive machines. Smaller still was the General, a Cleveland Tractor Company model, sold in selective markets by Massey-Harris.

After World War II the Model 20 was introduced using a Continental F124 engine, and 1946 saw a new 20/30hp Model 30 replace the 101 Junior. In 1948 the Pony was presented, the only tractor that Massey-Harris built in any number in Canada. It was the smallest in the Massey-Harris line-up and used a Continental four-cylinder engine, and was marketed in France as the Model 811, using a Simca engine. Later it was superseded by the 812 and 820 models, using

Alanson Harris was joined in partnership by his son John in 1863. The two worked together as the company merged and became Massey-Harris.

before, the tractor failed in its attempts to be everything to everybody. Although a very advanced machine for the time, with its permanent four-wheel drive, articulated chassis and 3703cc (226 cubic inch), four-cylinder, water-cooled Hercules engine, its four-wheel drive restricted the turning circle of the tractor and caused frustration to the farmer. The track could be adjusted but had to be pre-ordered, eliminating the possibility for the farmer to do it himself. A later update became the Four-Wheel Drive.

Massey Ferguson

When Massey-Harris bought the J. I. Case company in 1927, they also acquired the Wallis concern. This is the Wallis 20-30.

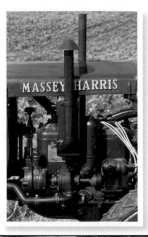

Initially and up to 1936, the Massey-Harris General Purpose tractor used a Hercules engine, which was then replaced with a company engine.

The Massey-Harris Number 3 was even larger than its predecessors, and used a Buda, four-cylinder, 6505cc (397 cubic inch) engine.

Hanomag, Simca and even Peugeot engines. The 101 was replaced by the Model 44 in 1947. A future success story for the company, this was a three/four plough machine and used the company's four-cylinder engine. For those who preferred a smoother feel, the 44-6 derivative used a Continental six-cylinder engine, and an LP gas model became popular with farmers as an alternative fuel model.

In 1946 the 55 model was the largest wheel-type farm tractor on the market, with a diesel version presented in 1949. Two more models followed in 1952 – the Colt and M23 Mustang. The Colt was only built for two years and used an F 124 Continental engine; the Mustang was a successor to the 22 model and used the same F 140 four-cylinder engine.

Back in the 1940s, Massey-Harris had turned down an offer from Harry Ferguson to produce his TO-20, equipped with all of his advanced attachments. They rejected the offer, and Ferguson took his tractor to America and struck a deal with Henry Ford, both companies then reaping the rewards. Massey-Harris knew that if they could not obtain a similar combination, they would be left

This is the Massey-Harris General Purpose tractor, which had four-wheel drive and an articulated body and very wide track.

The General Purpose Massey-Harris tractor was the first to be made and designed by the company, and although it had advanced features, it was not a great success.

For the mid-1930s, Massey-Harris produced the Pacemaker Tractor, which evolved from the first unit-frame, Wallis-based Massey-Harris tractor.

The 1936 Challenger model was a conventional row-crop tractor, and was also based on the original Wallis unit-frame construction.

lagging dangerously behind the competition. The new Ford management and Ferguson were going through a difficult time, and finally split, with Ford launching their Model 8N. Ferguson had built up a large distribution network in America, and he used this to sell his newly imported TE20 tractor, which had been launched in England back in 1946. Designated TO for America, he launched the tractor on the US market in 1948, kitted out with the complete Ferguson system. With this, and the launch of his new 30hp TO30 in 1951, he grabbed much of the Ford market. The production side of the business was not what Ferguson enjoyed, though, which resulted in him looking for someone to do that for him. How convenient that Massey-Harris should also be looking for exactly that kind of arrangement, and in this way a second bite of the apple presented itself, and the two

The Massey-Harris 101 Junior and 101 Super were introduced in 1939. The 101 Super (shown here) had a 3298cc (201.3 cubic inch) Chrysler motor, and both were available with standard front axle or row-crop tricycle, as seen here.

Derived from the 101 model, this Model 102 Junior also uses the six-cylinder engine fitted to its smaller brother.

Massey-Harris launched the little Pony 8/10hp model in 1947, but it was never a great success.

signed a deal in 1953, where Massey-Harris would buy out the Ferguson Tractor Company. The name was changed to Massey-Harris-Ferguson, which was later shortened simply to Massey-Ferguson (and eventually the hyphen was removed).

Initially, both brand names produced their tractors: the 44 became the 444 with extra power and dual-range transmission in 1955; the 33 also received a new transmission and became the 333; and the big 55 became the 555. The Ferguson TO-30 was upgraded to the TO35 and also marketed as the Massey-Harris 50, while the Ferguson F40 was offered in tricycle format. The new M-F tractors were now being painted in their new joint red and grey livery, with the Massey-Harris red and yellow livery relegated to the archives.

The M-F TO30 now brought up the bottom of the range, while the old Ferguson 35 was upgraded to the M-F 50 and a new M-F 65 was introduced. Some rebadging and buy-ins now happened, to see the company through a period of new development. For example, the Minneapolis-Moline 75hp Gvi model was given new bodywork, painted in the new colours and sold as the M-F 95

Super. The Oliver 990 model was rebadged, repainted and marketed as the M-F 98, and in 1959 M-F introduced their own large horsepower model, the 60hp M-F 88, a machine that used a 4522cc (276 cubic inch), Continental, four-cylinder engine. But with everybody screaming out for more power, M-F resorted to bringing in yet another Minneapolis-Moline and disguised it as the M-F97. In 1959 they bought the respected Perkins Diesel Engine Company of Peterborough, England, and now became builders as well as suppliers of engines.

Late in 1964, M-F introduced their next series of new small tractors, the Red Giants, as they were to be known. These would take the place of the now ageing 35, 40 and 50 models. The new 100 series machines consisted of the 135, 150 and 165. These were manufactured in Britain and France, while the larger M-F machines were assembled in the USA. At the other end of the scale was the new M-F 165 tractor, which replaced the M-F 65. With redesigned bodywork, the 165 was similar in many respects to its predecessor but did not remain the largest model for long, being replaced by the M-F 175. This used a larger Perkins engine and was replaced in 1971 by the M-F 178, with a larger 4063cc (248 cubic inch) Perkins diesel engine. The whole range

Production of the Model 55 petrol model began in 1946. This is a diesel version which was introduced three years later. Production ceased in 1955.

With its long engine cover and housing, the Mustang looked bigger than it was. It also lacked hydraulic, three-point hitch with draft control.

MASSEY FERGUSON

The Model 44 appeared in 1946 as a standard tread version, but was followed a year later by a row-crop version and a diesel standard model in 1948.

Massey Ferguson

Once the purchase of Ferguson was completed, one of the first tractors to be introduced with the Massey Ferguson name was the Model 65.

A huge tractor, this Massey Ferguson Model 95 is actually a Minneapolis-Moline Gvi in disguise.

This is the Model 97, and clearly viewed here at the front of the machine is the axle of the optional four-wheel drive it was equipped with.

was updated soon after with the designations 148, 168 and 188 – split into two versions: the budget series or standard rig models. The DX 1000 range saw the introduction of the M-F 1100 and M-F 1130. The biggest machine of this period, and only produced between 1970 to 1972, was the M-F 1150, using a Perkins V8 diesel engine, and for 1973 the M-F 1135 and M-F1155 were introduced for the American and Canadian markets.

1500 and 1800 four-wheel-drive models came in 1969 – this type of large machine was now more acceptable to the American farmer, who was seeing the benefits it could give. New 1505 and 1805 models offered more power, adjustable tread width and an optional rear PTO. The 1970s was the era of big and powerful, and in 1978 their Brantford, Ontario plant started producing the new four-wheel-drive 4000 series machines. These were the biggest machines they had ever produced, with the smallest model rated at 225hp and the biggest at 375hp. They

This is one of the last tractors to carry the Massey-Harris badge. This is an advertisement for the 555, an updated version of the 55.

Massey-Ferguson
MF97 TRACTOR

Massey Ferguson were looking for a large tractor, which their line-up lacked. They chose a Minneapolis-Moline and dressed it up as the 97.

The Massey Ferguson 135 had a long life. It was introduced in 1964 and was around for 11 years. It used a three-cylinder Perkins engine.

This Model 35 may look familiar. In fact, it has its roots with the Little Grey Fergie, and changed its designation when the companies merged.

all used the Cummins 14,797cc (903 cubic inch) V8 engine with various ratings, and all had 18 forward and three reverse gears.

The up-beat sales period of the 1970s was followed by a sudden reverse in fortunes during the 1980s, and M-F soon found the going hard. In 1988 the M-F 4000 series was replaced by just one very large model, the 5200. By now M-F were running short of operating capital, and in 1989 they sold the 5200 to McConnell Tractors Ltd. The design was advanced, but by 1995 McConnell decided it could no longer continue on its own and sold the line to AGCO. In 1993 AGCO acquired the rights to Massey Ferguson, and when in 1994 they needed a new four-wheel drive, they took the McConnell/M-F designed tractor to the AGCO/White, Cold Water factory for production. Here they upgraded it, and the new AGCO-Star was conceived, which had a choice of Detroit-Diesel or Cummins engine units. By 1973 the smaller series of tractors got a makeover and were redesignated as the 200 series, all being assembled at the Banner Lane works in England. Mid-range tractors were being

Massey Ferguson

This is the Massey Ferguson 650 turbo, a rugged machine that can tackle most rough terrains.

assembled in Beauvais, France, and the Detroit factory in the USA, while the larger machines were produced in Canada and the USA. The new M-F 500 model was released from the Banner Lane factory in 1976, and by the mid-1980s the cabs were painted all red to the roof, with a new grille incorporating the headlamps. The 1980s was a bad time for farmers in the USA, and M-F decided to close down their factory in Detroit – the one that Harry Ferguson had so hastily built so many years before, after having fallen out with the Ford Motor Company. The 500 series tractors were replaced by the 600 series, also made at Banner Lane, and the 200 series basically became the 300 series in 1986.

By the 1980s, the French arm of M-F in Beauvais were also making tractors in their own right, producing their new Topline 2000 models. These were medium-size machines and fitted nicely between the British-made smaller models and US and Canadian-made larger models.

Massey Ferguson was bought out by the AGCO Corporation, firstly their American section in 1991, and shortly after throughout the world in 1994. Once again, the company went through a rationalization period and large lines of tractors were slimmed down to make the company leaner.

An advertisement for the biggest tractor of its period of manufacture – the monster eight-cylinder Massey Ferguson Model 1150.

V-8 POWER
New MF 1150 V-8 Tractor

Manufactured between 1974 and 1977, the Model 1505 is a big machine that uses a Caterpillar eight-cylinder diesel engine.

The 300 range was topped by the Model 399. These were made between 1986 and 1997. They used a Perkins six-cylinder diesel engine.

For 2002, there was the new Massey Ferguson 4300 series, which set new standards in versatility and customization for the mid-range 55 to 99hp tractor market.

The 6400 series ranges from 120hp to the 215hp 6499 shown here. Introduced in 2008, the 6499 uses AGCO's SisuDiesel 74 engine and Dyna-6 transmission.

Two new series were introduced in 1995 – the 6100 and 8100 – and the Banner Lane works upgraded the old 300 series to the 4300 series in 1997. In 2001 the workers at the Coventry Lane works received further good news when the model was uprated once again, with a promise of a £1.7 million investment in the site. The company had dipped in and out of problems, with the 1,800 workforce already working short time. The spectre of closure was raised when AGCO warned that the company could pull out of Britain if it stayed out of the euro. The M-F 4200 tractors were sold throughout the world, and more than 90 per cent of Banner Lane production was exported globally. The strong UK currency now badly damaged sales – the euro gave a better exchange rate and was also being used all over the continent. Production stopped at the Banner Lane works, and the 4300 model became the last series produced. The closure of the factory in 2002, with the loss of a thousand jobs, was a major blow to manufacturing and the economy.

AGCO had decided to consolidate production in France and Brazil, and the Coventry office remained as the corporation's headquarters until 2006, when they were moved to nearby Abbey Park, Stoneleigh. The offices were finally demolished in 2012, bringing to an end an important part of the Massey Ferguson story.

The Beauvais plant is the most modern tractor facility in France, and is the centre of Massey Ferguson's Europe, Africa and Middle East operations. It is AGCO's largest European factory

and has produced more than 820,000 tractors, of which nearly 85 per cent have been exported to over 140 different countries.

In 2008, when the company celebrated the 50th anniversary of the first tractors with the Massey Ferguson name, they were producing 85 different

tractor models. Today, the M-F product line-up offers machines to suit every type of agricultural operation needed by the company's five million customers across the globe. The company is truly international in its breadth of experience and has a global network of over 3,000 dealers.

Machine of the year for 2004 was the Massey Ferguson 7400 series tractor. This is the 7480 model, which uses a Perkins six-cylinder diesel engine.

McCormick

The McCormick company has a long and illustrious history. Named after the man who initially started the company, Cyrus Hall McCormick, it was one of several that amalgamated back in 1902 to create the International Harvester Company.

The McCormick Model MTX BetaPower is able to take on the toughest of jobs with its four-speed powershift system.

The McCormick MC 100 has a flexible proven transmission with creeper option, which delivers the right speed for all operations, from planting to harvesting.

In late 1984 Tenneco, owner of the Case and David Brown brands, declared its intention to purchase certain assets of International Harvester's Agricultural Division, and a deal was concluded in 1985 in which International Harvester was placed under the control of Tenneco's Case division. From here onwards, all products from Case's agricultural division were rebranded as Case International.

In 1999 Case was looking for a merger with New Holland, and once the shareholders had agreed, the new company became Case New Holland Global N.V. (CNH). European Union regulatory authorities rubber-stamped the merger, providing that CNH divested themselves of the Doncaster Wheatley Hall Road plant and the 50 to 100hp C, CX and MXC tractors it produced, along with the MX Maxxum production and engineering know-how.

The McCormick CX model of 2006 was given a more efficient engine with power ranging from 60 to 83hp.

McCormick

Beginning in 2008, the new TTX range of tractors, from 190hp to the 230hp version shown here, were produced at the Fabbrico plant in Italy, after production moved there from the Doncaster factory.

In 2000, after negotiations with various interested parties, the plant was purchased by ARGO SpA of Italy, and it announced that the Doncaster plant would become the global headquarters for McCormick Tractors International Ltd. It was decided that products from Doncaster would be sold worldwide under the McCormick brand name. In January of the following year the EU authorities gave their approval, and McCormick Tractors International Ltd started trading.

In 2001 ARGO SpA bought the CNH transmission facility at St Dizier in France, and it became known as McCormick France headquarters. By 2002 McCormick distribution covered all of Europe, Australia, New Zealand and South Africa. McCormick USA was also set up to distribute products in the United States.

Production of the new McCormick range of CX and MC four-cylinder tractors began in 2001, followed by the MTX six-cylinder models. The line-up of tractors for 2005 included the C series,

McCormick also produce a series of crawlers. These are ideal for the rougher but more specialist duties.

The MTX 115hp to 195hp models feature powerful, environmentally friendly engines, a slick powershift, power shuttle transmission and highly efficient hydraulics and PTO.

CX range and the MC Power 6 series. The MTX range followed, as did a T series crawler. A V range tractor for vineyard work and the XTX range of tractors were introduced.

It was an XTX215 that was the last tractor produced at the Doncaster plant after production moved to Italy in 2007, with full production of the XTX and TTX tractors beginning there in 2008. That year also saw the first McCormick telescopic handler introduced.

Recent developments include the launch of the T-Max series in 2009, followed by new styling for much of the McCormick range, as well as the new X70 series with Tier 4 interim engines in 2011.

All in all, the latest McCormick range of versatile tractors can cater for all the jobs that need to be covered on the farm.

Minneapolis-Moline

Minneapolis-Moline was created in 1929 through the merger of the Moline Implement Company, the Minneapolis Threshing Machine Company, and the Minneapolis Steel and Machinery Company.

The letters MTMC, written on the side, give this machine away. It was constructed by the Minneapolis Threshing Machine Company.

Around 1886, Midwest American farmers were calling John Deere ploughs 'Moline plows', a name that John Deere had used to describe his own implements. Unfortunately, another local company also wanted to use that name for their products. Candee, Swan & Company produced a catalogue in which the non-John Deere products were almost mirrored, model for model, and which also included the Deere trademark, causing misunderstanding and confusion with local farmers. Following legal claims and counter-claims, and a change of names, the Moline Implement Company became part of Minneapolis-Moline. By 1929 the Minneapolis Threshing Machine Company and the Minneapolis Steel and Machinery Company had also joined the merger.

FACT BOX

Moline Implement Company

- **1867** Deere takes Candee, Swan & Co. to court.
- **1868** CS&C bought out and renamed The Moline Plow Company.
- **1869** November – case concludes with a victory for Deere & Company.
- **1871** Case appeals to Illinois State Supreme Court and the decision is reversed.
- **1915** Moline Plow Company buys the Universal Tractor Manufacturing Company of Columbus, Ohio, and starts production.
- **1920s** Moline Plow Company taken over by John N. Willys. Company changes name to the Moline Implement Company. Universal Tractors sold to International Harvester.
- **1929** Moline Implement Company merges to form the Minneapolis-Moline Power Implement Company.

The 21-32 Twin City was the last tractor model to use that badge. In 1929 the Minneapolis-Moline company was born.

FACT BOX

Minneapolis Threshing Machine Company

- **1887** The company is established as a threshing machines maker in Hopkins, Minnesota, and starts manufacturing steam traction engines.
- **1911** Walter I. McVicar designs the Minneapolis 35-70, which is followed by the 15-30 and others.
- **1928** Unable to continue alone, the company decides to join the merger.
- **1929** March 30 – the offer is accepted and the Minneapolis Threshing Machine Company becomes part of the new Minneapolis-Moline Power Implement Company.

A monster of a machine, this is the Minneapolis Twin City, six-cylinder 60-90. It had a massive engine capacity of 36,543cc (2230 cubic inches).

This is the Minneapolis Twin City Model 16-30. It used a water-cooled, four-cylinder engine, and looked more like a racing car than a tractor.

The Moline Universal Tractor was designed for use with either horse-drawn implements, or with the newly developed Moline implements.

FACT BOX

Minneapolis Steel and Machinery Company

- **1902** The company is founded in Minneapolis by J. L. Record and Otis Briggs. Its business is manufacturing steel components for the construction industry.
- **1910** Joy-Wilson Company of Minneapolis designs the Twin City 40. The largest Twin City is the 60-90. There is also a smaller 40-65. Building starts for the Case Threshing Machine Company and Bull Tractor Company.
- **1917** The 16-30 traditional tractor is presented. Unable to continue alone, merger negotiations start with the Moline Implement Company.
- **1929** March 30 – Minneapolis Steel and Machinery Company join the Minneapolis-Moline Power Implement Company to complete the merger.

The 1930s saw one of the greatest depressions of all time, but the three companies had done well to amalgamate their strengths. Most of the older tractors from the Minneapolis Threshing Machine Company and the Minneapolis Steel and Machinery Company were either phased out or were only produced for a few more years.

A Minneapolis-Moline Twin City KT (Kombination Tractor) with four-cylinder engine was introduced in 1929, and was then upgraded in 1934 to the KT-A. The company's first row-crop machine, the Universal 13-25, was introduced in 1931, and the Model J came in 1934.

1935 saw the updating of the FT and MT models, all of which used the Minneapolis-Moline four-cylinder engine unit – an A at the end of their designation denoted that. In 1936 the J was replaced by the Model Z, a four-cylinder, water-cooled machine with five-speed transmission – which was in itself a first.

For 1938 M-M presented the extraordinary UDLX (U-Deluxe) Comfortractor, which looked more like a car than a tractor. The enclosed bodywork was entered through a rear doorm and it had lights, a heater, a radio and even a passenger seat. Priced at around $2,000, it was too expensive for most farmers of the period, and only about 150 were ever made. Along with the U was the GT, which was made between 1938 and 1941 and used a four-cylinder engine, developing later into the GT-A.

When America entered World War II, M-M was required to do its bit for the war effort, but once the war was over, the company introduced new and updated tractors. The UTS was presented in 1948, the UTU row-crop came in the same year, and the UTC was presented in 1954.

Minneapolis-Moline

Is it a truck or is it a car? Neither! In fact, it is the Model UDLX Comfortractor. It was thought that farmers might take their family to town in this vehicle.

This is the Minneapolis-Moline Model R. It was the company's smallest tractor of the 1930s. It used a 2703cc (165 cubic inch) engine.

In 1951 M-M bought out the B. F. Avery Company of Louisville, which filled the small tractor slot for their product line. It was now that M-M introduced their rather unsuccessful Uni-Tractor, with a variety of specialist implements. The design was eventually sold to New Idea, who developed other larger machines from it.

The late 1950s and early 1960s saw M-M join in the power race, and also introduced power steering, torque amplifier, three-point hitch, live PTO and live multiple hydraulic outlets. The 335 – later renamed the Jet Star – and 445 – later renamed the Four Star – shared these additions when launched. Factory front-wheel assist came out in 1962, and fully articulated four-wheel-drive tractors were presented in 1969 – the 500 and 700

Introduced in 1937, the Model Z became popular from the start. The early Z had a 3047cc (186 cubic inch) engine, while the later ZA and ZB had a 3375cc (206 cubic inch) engine.

This is the standard-tread Model UTC with ultra-high clearance. This model used a water-cooled, four-cylinder engine.

The Minneapolis-Moline range of Model Gs was extensive and varied. They sold well, and lasted through to the 1960s.

Looking more like a complicated timepiece, this is the Uni-Tractor. The machine needed dedicated implements, which restricted sales.

The Four-Star model range was introduced in 1959, and lasted through to 1962. It used a four-cylinder Minneapolis-Moline petrol engine.

This is the biggest Minneapolis-Moline of the late 1960s/ early 1970s – the G1050 – which used a 8259cc (504 cubic inch) diesel engine.

range. M-M were also trading with Massey Ferguson, who were selling certain M-Ms – Gvi, G-705 and 706 – under their own brand name and colours.

The White Farm organization, having taken Cockshutt and Oliver under their wing in the 1960s, now took M-M over too. The mid-1960s were spent producing large tractors such as the G1000 row-crop, and variations followed, with much inter-company badge engineering being carried out. The first M-M Super Tractor, the A4T, came in 1969. It was assembled at the White factory and distributed by all three companies in their relevant colours. Even Fiat tractors were being rebadged and sold by M-M outlets – the G350 and G450 models. By the mid-1970s the only real M-M component worth a mention was its big six-cylinder engine, which White exploited to the limit. By now the M-M name was relegated to the also-rans, and only once more did it appear when the 1989 American 60 and 80 models were painted in the now-dying colours of yellow and white. After further corporate shuffles, M-M became part of the AGCO family.

New Holland

The New Holland name has been around for many years, but only since 1986 has it been seen written on the side of a tractor, subsequently growing into the brand leader that it is today.

In 2005 all the TG series new Holland tractors were powered by a six-cylinder, 8.3 litre, turbocharged engine.

Through mergers and acquisitions, this giant of a company is now part of the global CNH group. New Holland has more than 5,000 dealerships and distributors throughout the world, and is the leading international brand of agricultural machinery. They can offer farmers anywhere in the world exactly the right machine needed to cultivate, produce and grow, along with a full range of financial and after-sales services.

The New Holland tractor range is more than comprehensive, and caters for all requirements. It includes a selection of high, medium and low horsepower models, articulated frame tractors, bi-directional tractors, compact tractors for small-scale agriculture, special tractors for narrow, orchard and vineyard applications, crawlers and models created for home use and gardening. One tractor out of five currently at work anywhere in the world is a New Holland.

A closed centre hydraulic system, with an advanced load-sensing CCLS oil pump, is standard on all TVT tractors.

The latest T4 series has cabs that offer high levels of comfort and technology, including the ergonomic CommandArc console to the right of the driver.

All-purpose series TK80A, TK90A and TK100A tractors combine ease of operation and low operating costs with excellent stability and traction.

NEW HOLLAND

The 2006 series LM-A Telehandlers can be specified with a choice of 95 turbo, or 110hp turbo, intercooled engines.

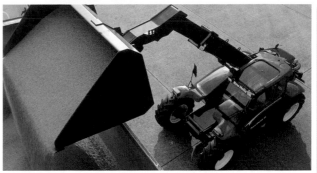

Today, for example, more than 10,000 New Holland grape harvesters are working in vineyards around the globe. New Holland offers grape harvesters to suit any type of vineyard. Building on its great experience and constant research, the multi-purpose models of the VL and VX ranges reflect New Holland's leadership in grape harvesters, with the latest addition of the Braud 9000L range cementing that reputation. To meet the diverse needs of the specialist customer, New Holland also offers the T4000 F/N/V ranges of compact, narrow-width orchard and vineyard tractors. New Holland also offers an industry-leading line-up of under 60hp tractors, including the T3000 series, which are ideal for homeowners and hobby farmers.

By offering a comprehensive range of products and services that provide farmers with highly effective production tools, New Holland are now the leading brand in the Latin American agricultural machinery market. When dealing with so many different crops, harvesting techniques and geographical differences, you have to be a specialist to serve all Europe-Africa-Asia region farmers and contractors. With innovations such as the 2009 presentation of the NH2, the world's first hydrogen-powered tractor and other design and technological advancements, New Holland is at the forefront of global agriculture.

New Holland's T4000 F/N/V series of tractors offers high standards for the specialist in need of narrow orchard and vineyard machines.

Oliver

In 1855 James Oliver of Mishiwaka, Indiana, USA, patented his Chilled Plow, which had a very hard outer skin. The plough became very popular and Oliver gained the title of 'Plowmaker for the World'.

It was 1870 when James A. Oliver sold his first 'chilled' plough, where the cast iron is cooled rapidly by water to harden the metal.

Oliver started experimenting with his own tractor in the 1920s, and produced a machine known as the Oliver Chilled Plow Tractor. Soon after, he agreed a merger with the Hart-Parr organization. After the death of Charles

FACT BOX

Hart-Parr firsts

- **1868** Charles H. Parr is born, USA.
- **1872** Charles Walter Hart is born, USA.
- **1892** Hart and Parr meet as mechanical engineers at the University of Wisconsin, USA.
  Their interest in producing an internal combustion engine together culminates in five being made. Both graduate with honors.
  A small factory in Madison is found to manufacture engines.
- **1900** First oil-cooled engine is produced.
- **1901** June 12 – New factory incorporated in Charles City, Iowa, USA.
- **1901** July 5 – Gasoline traction engine production begins, recognized as the birth of the farm tractor industry.
- **1902** Development of valve-in-head principle for tractor engines.
- **1902** July 19 – First gasoline traction engine is sold.
- **1904** First successful method of burning kerosene for fuel.
  First magneto ignition on tractor engine.
- **1905** First force-fed lubrication on tractor engine.
- **1918** The principle of outside counterweights on tractor engines is perfected.
  First kerosene fueliser is introduced.
- **1920** Alemite system of force-fed lubrication is adapted for tractors.

An early Hart-Parr machine, which was used mostly to work threshing machines of the period via its pulley and a leather belt.

A wonderful period photograph of an Oliver row-crop Model 70 being operated by a female farm worker.

The early Hart-Parr models were very heavy and slow. Soon they were being replaced by the lighter and more user-friendly 12/24E.

Parr – Charles Hart having left the company back in 1917 – the company merged with Oliver to become the Oliver Farm Equipment Company in 1929. The headquarters were established in Chicago, but the separate factories remained where they were. Oliver could now supply not only the tractors but also the implements to go with them too. The company changed its name to the Oliver Corporation in 1944, and purchased Cletrac, who had been around since 1916 as the Cleveland Motor Plow Company. They continued to produce crawlers until 1960, when Oliver was taken over by the White Motor Corporation.

After the merger of 1929, Oliver concentrated on producing new tractors, with the Oliver Hart-Parr 18–27hp row-crop machine arriving in 1930. This was followed by the 22–44hp Model A, a standard four-wheel machine, and with a boost in the power rating it became the 28–44hp, produced up to 1937 and the successor to the Hart-Parr 18-36. It used a four-cylinder, 443 CID,

Featuring both company names, this row-crop model was one of the first tractors produced after the two companies had amalgamated.

A full view of the Oliver row-crop Model 70, equipped with Oliver 'tip-toe' steel wheels. Rubber tyres became an option.

Oliver

1937 saw the introduction of the Oliver Model 80 row-crop. A further diesel version was presented in 1940.

Essentially a smaller version of the Model 70, this is the Model 60, which used a four-cylinder engine.

Production of the Oliver 70 series tractors finished in 1948. They were replaced by the Oliver 77, of which this is a Fleetline version.

kerosene-burning, valve-in-head engine, and was able to pull four- or five-bottom ploughs.

In 1935 Oliver moved away from the four-cylinder engine unit, producing a six-cylinder Waukesha engine, Model 70. The machine sold on its power and speed and certainly looked the part with its long nose, accommodating the water-cooled 3310cc (202 cubic inch) engine. The 70 was given Fleetline styling in 1937, and the smoother, rounder lines made it positively good-looking. Known in the USA as the Oliver 70 and painted green with red wheels, in Canada they were painted red with cream-coloured wheels and sold under the Cockshutt brand name as the 70.

For 1937 the older four-cylinder models were updated and given new designations. The 18-28 became the 80, and the 28-44 became the 90, and although they were older models, they became reliable workhorses for many farmers. The 80 Standard of 1940 was rated as a three-plough machine, and was available in both petrol and kerosene versions. The Model 90 used the familiar four-cylinder, 7259cc (443 cubic inch) engine, with a three-speed transmission. A more powerful Model 99 was introduced in 1937, and gained the distinction of having the longest production run of any Oliver tractor, being produced up to 1957.

For 1940 a miniature version of the 70 was produced, using a water-cooled, four-cylinder engine with four-speed transmission. The 60 was every bit as pretty as its larger stable-mate, and

In 1940 Oliver introduced the Model 60, which was replaced by the Oliver 66 in 1949. Production of the Oliver 66 tractors ended in 1952.

Seen here is an Oliver Model 99, which was made up to 1952, when it became the Super 99, and was often fitted with a supercharger.

This is the 1952 Super 99, which was a bit of a hot-rod machine. Its two-stroke engine was fitted with a supercharger.

filled a valuable model gap. By 1948 the 70 and 80 series were replaced by the 77 and 88 models. The 88 was new, and was fitted with the Waukesha six-cylinder engine, a six-speed transmission with two reverse gears, and it was given Fleetline styling. A year later, the 60 was given a power increase and redesignated the 66. The biggest machine in the Oliver stable was the Model 90, now upgraded to the 99 and fitted with a Waukesha six-cylinder, 4.9 litre diesel engine. For 1955 it was designated the Super 99, with a choice of Waukesha six-

This is a Model 66, powered by a diesel engine and with adjustable front-wheel track.

cylinder unit, or a General Motors, water-cooled, three-cylinder, supercharged two-stroke engine.

During the 1950s, the 66, 77, and 88 were once again upgraded to become the Super 66, 77 and 88, joined by two smaller models, the Super 44 and Super 55. Introduced in 1957, the Super 44 was produced at the Battle Creek plant in Michigan, and featured an off-set seating position, giving the operator better visibility of the work ahead. The Super 55 was produced from 1954 through to 1958 and featured a three-point hitch and a 144 CID, four-cylinder Oliver engine. As the 1950s came to

With the vast-acreage farms of the mid west, a sun canopy was something of a luxury. This is a Model 77 row-crop version.

Oliver

an end, all of these models were redesignated again, this time by adding a zero at the end of their respective numbers – so the 55 became the 550, and so on. There was extra power and slight changes in styling, and some had dual-range transmission, or Power Booster, as Oliver named it. The 770 model of 1958 was one example, and its six forward speeds were now 12. The six-cylinder engine was still there, but gained a power increase.

Numbers alone do not help with sales, and as the late 1950s turned into the early 1960s, Oliver was taken over by White Motors. Although not the end of the company, actual Oliver tractors would now be few and far between.

Other tractors were painted in the Oliver colours, the first coming from David Brown – the 500 and 600 models – painted in Oliver green. Two homemade models were seen in the early 1960s, the 1800 and 1900, followed in 1967 by the 2050 and 2150 turbocharged models. The 1250 of 1965 was a Fiat tractor, badged White-Oliver from 1969. The 1970s saw the 55 series, and many of the bigger machines were just repainted Minneapolis-Molines. By 1974 the end had come, and the last Oliver, a 2255, rolled off the Charles City production line.

Familiar tractors? These are repainted and rebadged David Brown machines – the 52hp Model 600, and a 32hp Model 500.

The Oliver 1855 of the early 1970s used a six-cylinder diesel engine, and had six forward and two reverse gears.

Built between 1969 and 1974, this is the Model 1955, which used a Waukesha turbo-diesel, six-cylinder power unit.

Renault

Louis Renault was born into a typically bourgeois Parisian family in February 1877. Although Renault automobiles would become a worldwide product, tractor production would soon follow too.

A wonderful period photograph taken in 1898, showing Louis Renault at his lathe in his workshop in Paris, France.

At an early age, Louis Renault developed an enthusiasm for all things mechanical, including engines and electricity. The Renault family had a second home in Billancourt, very near to Paris, and it was in a garden shed there that the young Louis set up his first workshop. At the age of 20 he converted his De Dion-Bouton tricycle into a small four-wheeled vehicle, and added another of his inventions that would soon propel the motor car into a new era – the 'direct drive', which was the first gearbox.

On December 24, 1898, Louis made a bet with friends that his vehicle could climb the 13-degree slope of the Rue Lepic in Montmartre. Not only did Louis win his bet, but he also pocketed his first 12 firm orders. His career was under-way, and a few months later he filed the patent for the direct drive system. However, it was through racing that Renault Brothers became known, with Louis and Marcel at the wheel of their vehicles. In 1909 Fernand died after a long illness, and so Louis was now alone, having lost Marcel in a tragic racing accident some years earlier. It was at this point that the company was renamed the Louis Renault Automobile Company.

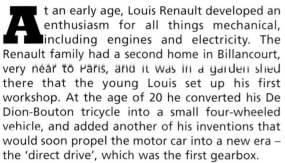

The Renault company is better known for its cars and trucks. They made their first tractor from a design derived from a tank. This is the H1 tractor.

A photograph taken in 1918 showing the assembly department for the Renault F17 tank, which was used to create the first Renault tractor.

Renault

By the late 1970s, the Renault tractor company was able to offer a range of engine sizes for their tractors. Shown here is a 40hp 421M model.

This Italian rural country scene of 2004 shows a Renault Model 65-34 taking its turn to collect the dried hay.

World War I broke out in 1914, and Renault factories switched from making cars to helping the war effort, producing all kinds of equipment – trucks, ambulances, stretchers and more than 8 million shells.

In 1917 Louis Renault designed the first light armoured tank, the famous FT 17, and it was only a couple of years after this, in 1919, that Renault produced their first tractor. Basically, this was derived from the FT 17 tank, but was used for agricultural work rather than military activites.

Once the armour plating was removed, the Renault PE tractor began to take the shape of a tractor rather than a tank.

This was followed by a new machine that did away with the armour plating, therefore reducing the weight and increasing the field manoeuvrability. The Type HI was shortly followed by the HO, which was a wheeled version that was much more advanced than its predecessors.

In 1926 the PE model was presented, which was Renault's first purpose-built wheeled tractor. It used a 20hp petrol engine, and improvements were added as the years went by. It was converted to run on producer gas, as petrol prices became more expensive in the 1930s. As Renault had been experimenting with diesel engines since 1929, it was not a surprise to see that they introduced the YL tractor in 1933. This used a

Deep in the mud, this Renault Ceres 340X has no problem getting grip. The TZ model cab was one of the most advanced of its kind.

30hp diesel engine and featured rubber agricultural tyres. An alternative petrol engine version was also available – the VY.

At the start of World War II the company was once again obliged to produce materials for the war effort, and although they ceased making tractors, development continued. For 1941 the 301 model was presented in diesel and producer gas variations, and the 306 and 307 diesel and petrol models came next. In 1947 the company launched the R3040 series, the first production tractor to feature a complete electrical system, and also the first machine painted in the familiar orange livery.

The old Renault factory at Billancourt was on an island in the middle of the river Seine, where expansion was impossible. In early 1950 Regie Renault, as they were now known, and under state ownership, invested in a factory at Le Mans, with a view to increasing production capacity. The new factory would be well served by rail, and included a foundry, transmission machining and assembly department and final assembly line.

For 1956 a new tractor was introduced, the Series D, which had a choice of water- or air-cooled diesel engines.

In 1963 the company presented a 55hp model, which featured a 12-speed gearbox for the first time. Just two years later, the Super D series was introduced, which incorporated all these features along with tracto-control and direct-injection diesel engines. 1967 saw the introduction of the first four-wheel-drive tractor, and a year later Renault introduced a specialist vineyard and

The Tonka-style Ares 500 series tractors from Renault were packed with modern electronics. Today, Renault tractors are part of the CLAAS organization, and the Ares name is still used.

Looking like some strange multi-legged animal, this is the powerful top-of-the-range Renault Atlas model.

A bird's-eye view of the interior of the 1987 TZ model cab. This was a state-of-the-art workspace, and everything was within reach.

orchard model for the first time. The Renault tractor division was given its own identity in 1969 when it became Renault Motoculture.

In 1974 they introduced 40 new tractors rated between 30 and 115hp, all equipped with technological advances such as shuttle transmission, and available in two- and four-wheel-drive versions. For 1980 Renault Motoculture became Renault Agriculture, with their own unique logo. In 1981 the new TX style cab was presented for models between 83 and 135hp, which gave the operator more comfort and better visibility, but for the economy-conscious farmer the basic TS cab was also available.

For 1987 all the lower-rated tractors were given a facelift and renamed the LS series, while the 65 and 75hp models, using Perkins engines, became the SP models. These were followed by the MX and PX tractors, which had revised cab controls and featured a new gearbox and improved steering. During 1993 Renault introduced their Ceres tractors, with sloping bonnets and a choice of two cabs. The following year an agreement was signed with John Deere for Renault engines to be sold by them, and for John Deere to sell Renault tractors. An agreement was also signed with Massey Ferguson for a joint venture for the design, development and manufacture of drive lines. In 1997 the Ares and Cergos tractors made their debut, and by 1998 Renault were producing 9,000 tractors per year.

A powerful tractor was presented for the late 1990s in the Model 155.54 Renault. This was equipped with a turbocharged engine.

Small and manoeuvrable, this is the little Fructus 140 model Renault. This was aimed at the small farmer and for orchard work.

In spring 2003 CLAAS acquired a controlling interest in Renault Agriculture, and bought the remainder in 2008, making it completely CLAAS-owned. A range of industry-leading tractors now wear the CLAAS badge and colours.

The 2006 Axion tractor series from CLAAS offered between 163 and 225hp rated power output. It also incorporated the new Cebis terminal, new multi-function armrest and unique four-point cab suspension. In addition, the Axion came equipped with the convenient Hexashift transmission and innovative Drivestick control.

SAME

The SAME (Società Accomandita Motori Endotermici) company was founded in Treviglio, Bergamo, Italy, by Eugenio and Francesco Cassani in 1942. World War II was raging, and there was a lack of raw materials and few production facilities available.

The first diesel tractor, designed and produced by the engineer Francesco Cassani. It was a two-stroke, two-cylinder machine.

By the end of the 1920s, Francesco Cassani, with help from his brother Eugenio and huge support from his mother Luigia, and after many frustrating and sleepless nights, designed a tractor that was equipped with a diesel oil engine. Unfortunately, due to problems with distribution and manufacture, Francesco was beaten to the line by the Americans, who produced their own Caterpillar Model 65 – the first commercial fuel-oil tractor.

By the end of the 1920s, the tractor market had taken off in Italy and the Italian government, with its fascist regime, launched a competition to find the best Italian tractor. With such illustrious names as Fiat, Landini and Motomeccanica taking part, it was surprising that the little Cassani came out on top. With this, a search was started to see who could help to manufacture the machine. One approach came from Breda, but Francesco refused and decided to go with the Barbieri company. Although they seemed like a good, reliable company, this was not the case. They were already having financial problems and were hoping that the Cassani deal would help to get them out of trouble. Things went badly, and finally, after surrendering his share, designs and patents, Cassani was saved from going bankrupt, being dissolved by agreement of the partners.

Throughout the 1930s, Francesco designed diesel oil engines for trucks, boats and aircraft. During this same period, due to the lack of availability, he designed his own injection pump, and founded Spica – which stood for Società

The SAME Iron range are multi-functional tractors that are specially suited to medium to large farms and contract operators. They are equipped with a reliable Deutz four- or six-cylinder engine.

SAME

This SAME D30 diesel model was a step forward from the D25. By this time, the company was experiencing good sales with these tractors.

The SAME 240DT of 1958. This was equipped with a two-cylinder, air-cooled engine, producing 42hp. It had five forward gears and one reverse gear.

Pompe Iniezione Cassani – a company that specialized in the manufacture of injection apparatus for diesel engines.

World War II left Italy devastated, with both Allies and enemy leaving behind a country so ravaged that it struggled to survive. It was at this time that Francesco designed a motor-reaper – a kind of tricycle with an 8hp engine. Cassani had viewed this as a tractor that could be coupled with a variety of agricultural implements.

After a brief visit to England, Francesco was convinced that the agricultural mechanization he had witnessed should also be applied in Italy. On his return he designed a new and economical tractor – the first to bear the name SAME – a 10hp three-wheeler with a reversible seat. This became known as the Trattorino universale (all-purpose small tractor). Not too many farmers could afford a new tractor, though, and when they could, it would probably be a Ford, a Ferguson, or possibly even a Fiat, but they were unlikely to go for a tractor of an unknown make.

The Iron model uses an electronically controlled lift system with a capacity of up to 6,200kg (13,668lb), and is equipped with an oscillating damping system to stop heavy implements pitching.

One area he wanted to develop for his machines was four-wheel drive. Certain that this would put him ahead of his rivals, he set about producing one, which he did in 1952. The DA25 was the first diesel four-wheel-drive tractor, and without fail, sales started to pick up. Although the factory was small, it grew quickly, and by 1955 it was selling 1,750 units. More space was now needed, and Cassani took over the Caproni factory. The twin-cylinder DA38 and three-cylinder DA47 Supercassani followed, as did the Sametto, a machine that became the parent of a long and illustrious family.

By now SAME were becoming known worldwide, and the first dealer able to assemble the four-wheel-drive tractors was opened in Albertville, France. A new factory was completed in 1957 and production reached 3,000 units, but Francesco wanted more. Another trip to South America was made in 1959, unfortunately cut short due to the death of Eugenio. New engine designs followed, and a compact V configuration engine was developed. In 1958 the Automatic

The 2005 SAME Silver 110, shown above, was equipped with a new cab incorporating a high visability roof. It used the SAME Euro II, series 1000, naturally aspirated or turbocharged engine.

Easy-to-reach controls are situated on the right of the operator's seat. The 110 is equipped with an Agroshift gearbox featuring 60 forward and 60 reverse gears, with creeper function.

The Minitauro used an air-cooled, three-cylinder SAME engine. It was available with two- and four-wheel drive.

Linkage Control Unit was fitted to the twin-cylinder 240, soon to be followed by the 360 three-cylinder models.

On March 3, 1961, the new Samecar was presented, a kind of dual-purpose tractor with a truck-like cabin at the front. Although it was actually a sound idea, the time for it was wrong and production was stopped, and the Samecar was forgotten. The 60hp Centauro, using a SAME four-cylinder, 3400cc (207 cubic inch), air-cooled turbo engine was presented in 1965. The Leone 70 and Minotauro 55, two tractors that established the SAME name throughout the world, were also presented. In 1968 the company launched the Drago 80hp tractor, which used an L series, in-line, six-cylinder engine.

In 2005 SAME introduced the Argon Classic, a small tractor ideal for row-crop work on small farms.

While on his travels in Bolivia, Francesco got his ideas for the new Dinosaur model, which was to use a V8 engine, but unfortunately after several prototypes were made, it was abandoned.

It was not long after this that Francesco fell ill, and it was only a matter of time before he would no longer head the company. After his death in 1973, Hurlimann of Switzerland were bought, but the good times were coming to an end, with a world agricultural recession. Demand fell, and the company started to feel the pinch. There was now a period of managerial change and change again. New working methods, stricter work practices and modernization all helped to keep the company ahead and in good stead for the future. The Panther model was launched in 1973, and two years later the Tiger 100 was launched and awarded the 'best tractor in Europe' tag. For 1983 the Laser and Explorer families were launched, and the end of the 1980s saw the electronic era begin for SAME, with new electronic control units for various functions installed on the tractor. The Titan series was launched in 1991, and by 1993 a smaller series of tractors was designed for traditional applications.

Deutz-Fahr came on the market in 1996, and SAME, SLH Group as they were now known, purchased the company and put it back into profit by 1998. Today the SAME Deutz-Fahr group have a selection of tractors to be proud of, including the latest Vitrus range with Tier 4 compliant emissions. In 2011 they produced their one-millionth tractor – a vinyeard specialist Frutteto3.

The SAME 100.6 was introduced between 1995 and 2004. It used a six-cylinder diesel engine.

One of the larger models for 2005 was the Silver 85, which used a four-cylinder, turbo-diesel engine.

Steyr

The Steyr company was established in 1864 when Josef Werndl launched his arms factory in the town of Steyr, Austria. It was much later, in 1947, when the company moved into serious tractor production.

This Steyr 8080 is quite a significant tractor – it uses a bio-diesel mixture fuel that is derived from rapeseed.

The first Steyr machine was the Model 180 of 1947, a water-cooled, two-cylinder diesel, which produced 26hp. This was made up until 1962, and was manufactured using common components from the parent company's truck division. It sold some 45,000 units. Two years later, in 1949, the Steyr 15hp Model 80 was introduced; this was a single-cylinder machine, with around 65,000 units being sold up to 1966. Austrian farmers now wanted more out of their tractors, and Steyr responded by introducing their Jubilee series in 1964. These had hydraulic draft control, dual-speed PTOs and four-wheel drive, and were the first to be specially designed for front loader operation. 1967 saw the Plus series being manufactured, and these came in a choice of two-, three- and four-cylinder models – the 430, 540 and 650, which had 34hp, 40hp and 52hp engines respectively. These tractors went on to sell 110,000 units before they were phased out in the 1980s.

Steyr philosophy was to carry out constant development, and 1972 saw the Austrian farmer presented with a gearbox that had 16 forward and six reverse gears. In 1974 the Series 80 range of machines was introduced, ranging from 48 to 165hp, using a unique flat-floor cabin, which allowed the operator's seat to be positioned better and allowed the driver to sit more comfortably, while the hand-operated levers could also be better positioned. In 1975 the Steyr 1400a was introduced, producing 140hp and

Introduced between 1947 and 1953, the Steyr 180 model uses a two-cylinder engine. This is a rare 1952 model.

This 2005 CVT model tractor, along with the rest of the model range, was powered by a 6.6 litre, turbocharged, low-emission diesel engine.

Steyr

The CVT 6190 was the top-of-the-model range for 2005. In the following year it was upgraded to the 6195.

The popular Steyr Kompact range for 2013 is available with 65, 77, 86 or 97hp from the four-cyclinder, turbocharged engine.

The hitch and remote valves on this 4110 Profi CVT of 2013 are powered by a state-of-the-art closed centre, load sensing (CCLS) hydraulic system.

fitted with a turbocharger, powershift transmission, four-wheel drive and soundproofed cab. For 1980 Steyr was offering their bi-directional, 260hp, 8300 model, while their 48hp, three-cylinder, 8055 model and their two four-cylinder models were selling well. New developments were added to the tractors in 1986: SHR, an adjustable electronic hitch control system, and Informat, an electronic management system.

The 900 series of 1992 offered high PTO outputs and were designed to negotiate the hilly terrain in Austria. For 1993 the 9100 series and 9000 series tractors were presented, and a year later the Miltitrac upgrades featured sophisticated, multi-speed PTO controls and state-of-the-art linkage mechanisms. The M900 series, rated between 65 and 75hp, and the larger M9000 series was also launched, the range being expanded in 1995.

In 1996 Case IH took a controlling interest in Steyr, enticed by the work they had done on their front axle and transmission technology. This was further developed in 2000 with a Steyr-designed prototype transmission, used by the Steyr CVT and Case IH CVX models. In 2006 the Case IH and Steyr European Headquaters were opened in St. Valentin in Steyr's home country of Austria. The following year they celebrated 60 years of tractor production and brought out the 9000MT series.

Their continuing research and development resulted in the introduction of the CVT ecotech series in 2010, with an innovative gearbox design and continuously variable transmission, which fulfils European Tier4a emission standards. Steyr have always looked to the future, and the stability of being part of Case has meant that they continue to expand, with an excellent range of tractors.

Automatic four-wheel drive and differential lock are standard equipment on the CVT, and so they can tackle snow clearance with ease.

Valmet-Valtra

The very first Valmet tractors were manufactured in 1951. Today, Valtra tractors are built in the most advanced factories in the industry at Suolahti, Finland, and Mogi das Cruzes, Brazil. Valtra is a worldwide brand of the AGCO Corporation.

Launched in 1954, this is the Valmet 565 model, which had synchromesh transmission and a 52hp engine.

When Valmet introduced their 15A model, it was at a time when small tractors were replacing horses as the principal means of power on the farm. It weighed 780kg (1,719lb) and was powered by a 15hp, 1.5 litre, four-cylinder petrol engine, which was manufactured by the Linnavuori engine factory. Sales got off to a good, profitable start, and by September of 1954, the 2,000th Valmet tractor was delivered to its owner. The Valmet 20 was introduced in May 1955 to respond for the quest for more power. This was also powered by a petrol engine, but now produced 22hp, and could be fitted with a hydraulic lift linkage.

The Valmet 1502 is without doubt a strange-looking tractor. It was fitted with a six-cylinder, turbo-diesel, 6636cc (405 cubic inch) engine, and was made between 1975 and 1980. There were 16 forward and four reverse gears.

It was Gustaf av Wrede, the Director General of Valmet Oy, who instigated the development of what was later known as the Valmet 33, a tractor designed at the Tourula factories in Jyväskylä, under the leadership of Olavi Sipilä, who had also been instrumental in the design of the smaller Valmet tractors. The new tractor was launched at the Helsinki Messuhalli Fair in November 1956. By 1958, Valmet had begun exporting tractors to Brazil and China. Exports to Brazil started promisingly, but the Brazilian government decided to commence domestic tractor production almost immediately. Competitive bids were

The very popular Model 361D was launched in Finland in 1960, and had a 46hp engine.

147

Valmet-Valtra

The Valtra Valmet 8550, part of the 8000 family, was made between 1995 and 2003. A six-cylinder, 160hp engine was used, and 36 forward with 36 reverse gears were made available.

Valtra have had a presence in Brazil for many years now. This is the big 8550 model during the sugar beet harvest.

invited, and of the ten applicants, six, including Valmet, ultimately commenced production. In 1960 a new factory in the town of Mogi das Cruzes, near São Paulo, was commissioned, and the first 40hp Valmet 360D tractors were completed. Initially, Valmet had to use MWM power units, but they later fitted their own diesel engines.

Back in Finland, Valmet responded to the request for further power and launched its largest tractor so far, the 864, in 1964.

Olavi Sipilä moved to Tampere at this time, and Rauno Bergius was assigned the job of product development manager at the Tourula unit, with his first design being the Valmet 565.

Tractor cabs were starting to come under scrutiny for safety, and a frame was to become an obligatory feature by law. The Valmet 900, introduced in 1967, was the first tractor which used an integrated safety cab as standard. With tractors now having cabs, a taller assembly area was also needed, and so a new factory had to be located. The initial location was going to be near Jyväskylä, but an alternative place was found 45km (28 miles) away, at Suolahti, where production commenced in September 1969.

In 1979 Valmet partnered with Volvo and produced tractors under the Volvo BM Valmet brand. This is the turbo-diesel Model 2105 made between 1986 and 1987.

Valmet-Valtra

For 1999, Valtra produced this little Model 3500 vineyard tractor. It is ideal for grape collection.

The A series range of 50 to 100hp tractors are the most popular in Scandinavia, and include orchard specialist versions. This 2012 model has the unique mid-frame fuel tank for high ground clearance.

A Valtra T series tractor from 2005. This well-balanced vehicle is spreading its weight by having eight wheels.

Valtra launched the new T series in 2002, the same year that the Kone Corporation acquired the Partek Corporation, who owned Valtra. As of January 5, 2004, Valtra became a part of the AGCO Corporation, after being sold once again.

Valmet now introduced the turbocharged, 115hp, four-wheel drive, 1100 model. Ergonomic features were further developed, and in 1971 the company introduced the Valmet 502, which claimed to have the quietest cab in the world. In 1973 Valmet became market leader in Finland, and in 1975 introduced the six-wheel, 136hp, 1502.

In 1978 Valmet introduced the four-wheel-drive 702-4 and 702S-4 models, which had an impressive ground clearance of 47cm (18in).

The Valtra name had been registered back in 1963, and in 1970 it was used to brand a range of implements engineered specifically for operation with Valmet tractors.

Following negotiations with Valmet, Volvo BM made a strategic decision to cease manufacturing tractors and farm machinery. However, the company wanted to remain a component supplier, so the tractor operations of Volvo BM were transferred to Scantrac, which was 50 per cent owned by Valmet. By 1979, the design of a new range of tractors, designated Volvo BM Valmet, had already started. In 1982 Valmet introduced the red 04 models, ranging from 49 to 67hp, and a new 05 line, ranging from 65 to 95hp. In 1983 Volvo BM Valmet tractors became market leaders in all the Nordic countries.

In 1986 Valmet and Steyr-Daimler-Puch AG manufactured a range of engines, and also a series of tractors ranging from 90 to 140hp. In 1989 Valmet withdrew, but based on the knowledge gained from the project, they developed the Mezzo and Mega ranges. With a number of advanced features, the four-cylinder Mezzo tractors, introduced in 1991, and the later six-cylinder Mega models, enjoyed rapid sales.

The 1990s was a turbulent decade for the company, with ownership changing several times. Valmet Oy wanted to concentrate on their paper manufacturing machinery; as a result, Sisu acquired the tractor division in 1994. Valmet, as a division of Sisu, became part of the Partek Group in 1997, in turn acquired by The Kone Corporation in 2002. Kone sold Valtra to the AGCO Corporation in 2004.

Under AGCO, the brand has continued to expand through its facilities in Finland, as well as Brazil, helping to make it the second most popular tractor in South America. The company celebrated its 60th year in 2011 and currently offers a full range of tractors, from the A series starting at 74hp to the maximum-power S series at 370hp. Among Valtra's unique features is the TwinTrac reverse-drive system, which allows the driver to better control the tractor in reverse by means of a pivoting seat and secondary controls positioned at the rear of the cab.

White

Of all the Cleveland-based auto manufacturers, the longest-lived is White Motor, which started by making steam-powered cars. It also became one of America's major truck manufacturers, while also producing agricultural tractors.

The White 'American' models were introduced during 1989, and there were both 60 and 80hp versions.

Thomas White of Massachusetts, USA, founded the company and moved to Cleveland in 1876. They entered the tractor business by buying up companies that were struggling to stay afloat in the 1960s. The first to be bought was Oliver in 1960, although production of the 55 model continued. White also purchased the Cockshutt Farm Equipment Company in 1962, and then acquired Minneapolis-Moline a year later. The brand names of all three companies were retained by White until 1969, when the entire company was restructured as the White Farm Equipment Company, with its headquarters in Oak Brook, Illinois. The headquarters of the parent company, White Motor Corporation, remained in Cleveland, Ohio. By now the merged companies were mixing and matching components, which obviously made good

Sold as a White in Canada but as the Minneapolis-Moline in the USA, this is the four-wheel-drive Plainsman model.

By the time that the 2-70 Field Boss was introduced, petrol-powered tractors were losing ground to the diesel versions.

White

This is the White 2-60 model, which was in fact a Fiat in disguise. It was a replacement for the mid-range Oliver.

When White took over Minneapolis-Moline, things started to get confusing, as can be seen with the names on this tractor.

economic sense, but it aggravated tractor enthusiasts. It was in 1974 that all machines used the White name, although most were still being produced in their respective factories, and still using their own components.

The first articulated, four-wheel-drive tractors of the White brand name were presented in 1969. The new 139hp machine shared by all three companies was known as the Oliver 2455, the Minneapolis-Moline A4T-1400 and the White Plainsman A4T-1600. The following year, the company introduced the 6 series Oliver 2655 and Minneapolis-Moline A4T-1600; all used a 9586cc (585 cubic inch) engine, rated at 169hp.

White's advertising of 1974 stated, "like nothing you've ever seen before", and the new White 4-150 Field Boss was introduced, powered by a Caterpillar V8 engine. The White 'Boss' line styling and colour now replaced the clover green, prairie gold and white sumac red of the older models. In 1975 it was joined by the 180hp 4-180 Field Boss, a new and powerful four-wheel-drive machine that offered air conditioning, a heater and a protective roll-over cab, and in 1978 the 210hp 4-210 joined the Boss line. The dashboard had a 14-channel monitoring display, and night visibility was enhanced with two extra headlights placed on top of the grille.

While all of this progress was being made, the company found itself sinking deeper into financial problems, and in 1976 White Motor Corporation and Consolidated Industries planned to merge.

The Model 60 'American' was the smallest of the range. It was also the smallest tractor being made in the USA at the time.

White

The AGCO-Star shared its heritage from AGCO's three main tractor lines. In 1994 AGCO purchased the McConnell-Marc four-wheel-drive line, once Massey Ferguson. AGCO used the design to create the AGCOStar, which was painted in White silver; its decals bore Allis orange, Massey Ferguson red and White silver pinstripes. These were produced at the AGCO-White plant in Coldwater, Ohio.

The engine of the 2-180 was an eight-cylinder Caterpillar diesel of 636 cubic inches. There were 18 forward and six reverse gears.

Unfortunately, this did not work out, and the future of White Motors was left in doubt. In 1977 Consolidated Freightways Incorporated took on the distribution of the Freightliner trucks, and White lost 40 per cent of its truck sales. By 1980 White had run out of cash and were obliged to file for bankruptcy. In November they announced they were going to sell the farm equipment line to TIC Investment Corporation of Dallas, Texas, who resumed production the following year under the name of WFE – White Farm Equipment.

In 1982 WFE informed its dealers that two new four-wheel-drive models were on the way. One of the new tractors was the 225hp 4-225, which replaced the 4-175 and 4-210, but its styling and configuration were very similar. The wheel tread was adjustable and the machine could be used for row-crop work. The other machine, the 4-270, had bigger tyres and a larger articulated frame – and still used the Caterpillar V8 engine. The WFE products were made between 1982 and 1988, when the line once again found itself with financial problems. TIC offloaded the division to Allied Products, who already owned the New Idea

Equipment line, and the company became known as White-New Idea. Five new tractors were introduced, with engine power ranging from 94–188hp. Power units came from CDC diesel, a joint venture between Case and Cummins.

AGCO now stepped in and purchased the White and Hesston brands, before also taking over the Massey Ferguson concern in 1993. What they lacked, though, was a good four-wheel-drive machine for their large tractor line. The McConnell-Marc, originally a Massey Ferguson design, was a good starter, and so it was moved to the AGCO/White Cold Water, Ohio, tractor plant for production. Once installed, it was fine-tuned and kitted out with the latest gadgetry, given the name AGCO-Star, and introduced in 1994. In order to keep everybody happy, the machines were painted in the White colours – white and silver – and carried an orange Allis badge. In 2001 AGCO acquired the Caterpillar Challenger tracked vehicle line, and the large four-wheel-drive AGCO-Star models were pensioned off.

The White 6065 was a Lamborghini in disguise. The White brand was dropped in 1991 after the company was taken over by the AGCO corporation.

At the end of their life, White, or WFE as they were then known, were using disguised Lamborghini tractors, such as this 6105.

The massive V8 Caterpillar engine can be easily spotted under the long bonnet of this 4-150 White tractor.

Timeline

18th century — Oxen and horses are used for power. There are crude wooden ploughs, but all sowing is done by hand, cultivating by hoe, hay and grain cutting with a sickle, and threshing with a flail.

1830 — About 250–300 labour hours required to produce 100 bushels (5 acres) of wheat with walking plough, brush harrow, hand delivery of seed, sickle and flail.

1834 — McCormick reaper patented.

1837 — John Deere and Leonard Andrus begin manufacturing steel ploughs.

1850s — M. and J. Rumely Co. formed.

1855 — James Oliver invents chilled steel plough.

1868 — Deere & Co. incorporated.

1876 — First Case steam engine.

Brantford Plow Works.

1879 — McCormick Harvesting Machine Co. incorporated.

1882 — Cockshutt Plow Co. incorporated.

1890 — 35–40 labour hours required to produce 100 bushels (2½ acres) of corn with 2-bottom gang plough, disk and peg-tooth harrow, and 2-row planter.

1891 — Massey Mfg. Co & A. Harris, Son & Co. merge in Ontario to become Canada's largest agricultural firm.

1892 — Rudolph Diesel develops new engine while still a student.

Froelich builds his first gasoline engine at Froehlich, Clayton County, Iowa, USA.

1897 — Charles Hart and Charles Parr form Hart-Parr Gasoline Engine Co.

1902 — Hart-Parr makes its first gasoline-traction engine. Hart and Parr are credited with coining the term 'tractor' for the traction engine.

Integration of McCormick Harvesting Machine, Deering Harvester, Plano Mfg. Co., Milwaukee Harvester Co. and Champion Reaper Works into International Harvester.

Wallis Tractor Co. manufactures the Wallis Bear gas tractor.

1904 — Holt Mfg. Co. sells first steam-powered tracked tractor.

1915 Advance-Rumely formed from Gaar-Scott & Co., Advance Thresher Co., and Northwest Thresher Co.

1916 Cleveland Motor Plow Co. formed.

1917 Henry Ford begins manufacture of Fordson tractor.

1906 Holt makes first gasoline-powered tracked tractor.

1908 Heider manufacturing Co. manufacture their first tractor.

First Deutz tractor introduced.

1910 Rumely Oilpull begins production.

Landini introduce their first tractor.

1928 Massey-Harris buys J. I. Case Plow Works and with it the Wallis Tractor line.

Fordson production moves to Ireland.

Holt registers 'Caterpillar' trademark for tractor line.

Best forms C. L. Best Gas Traction Co.

1911 First Case gasoline tractor.

Minneapolis Threshing Machine Co. develops its line of gasoline tractors.

1914 Allis-Chalmers introduces first tractor.

Minnesota Steel & Machinery develops its line of Twin City tractors.

1918 Deere buys Waterloo Gasoline Traction Engine Co.

1919 Moline Plow Co. develops Moline Universal tractor.

Wallis Tractor Co. merges with Case Plow Works.

Fiat introduce their first tractor – Model 702.

1920 University of Nebraska begins tractor testing with John Deere Waterloo Boy Model N.

1921 McCormick-Deering 15-30 Gear Drive tractor replaces Titan tractor.

1923 Production of Deere Model D 'Johnny Popper' begins.

1924 IH introduces Farmall tractor.

1925 Holt & C. L. Best combine to form the Caterpillar Tractor Co.

1929 Oliver Tractor Company formed from Hart-Parr Co., Oliver Chilled Plow Co., Nichols & Shepard, and American Seeding Co.

Allis-Chalmers changes paint scheme from green to Persian Orange.

Timeline

1929 (cont.) Minneapolis-Moline Power Implement Co. formed from Moline Implement Co., Minneapolis Steel & Machinery Co., and Minneapolis Threshing Machine Co.

1930 15–20 labour hours required to produce 100 bushels (2½ acres) of corn with 2-bottom gang plough, 2m (7ft) tandem disk, 4-section harrow, and 2-row planters, cultivators and pickers.

15–20 labour hours required to produce 100 bushels (5 acres) of wheat with 3-bottom gang plough, tractor, 3m (10ft) tandem disk, harrow, 3.6m (12ft) combine and trucks.

1931 Caterpillar manufacture a crawler tractor with a diesel engine.

Allis-Chalmers take over Advance Rumely Thresher Co.

1932 Allis-Chalmers shows Model U equipped with rubber pneumatic tyres developed with Harvey Firestone.

1933 Fordson production moves to England.

1934 Harry Ferguson develops his innovative hydraulic draught control system.

IH introduces the W 40, the first six-cylinder IH tractor, and the first wheel-type tractor to be powered by a diesel WD-40.

David Brown introduces the Ferguson Brown tractor.

1937 Minneapolis-Moline adopts Prarie Gold colour scheme.

1938 Minneapolis-Moline introduces Comfortractor.

1939 Ford introduces Model 9N, the first tractor with Ferguson hydraulically operated 3-point hitch system.

Raymond Loewy restyles the new streamlined Farmall tractors.

1941 Minneapolis-Moline introduces first LPG-powered tractors.

1944 Oliver acquires Cleveland Tractor Co., and adds Cletrac tracked models to their line-up.

1946 Oliver introduces live PTO, an industry first developed by Cockshutt – Model 30.

1947 Ferguson sues Ford for patent infringement.

Nuffield produces their first tractors – M3 and M4.

1953 M-H merges with Harry Ferguson to become Massey-Harris-Ferguson, then later to become Massey Ferguson.

1955 6–12 labour hours required to produce 100 bushels (5 acres) of wheat with a tractor.

1959 Massey Ferguson buys diesel engine maker Perkins.

1962 White Motor corporation buys Cockshutt Farm Machinery.

1963 White Motor corporation buys Minneapolis-Moline.

1967 Case's majority stockholder is bought by Tenneco.

1981 White Farm Equipment is sold and continues operation as WFE.

1984 International Harvester farm equipment line bought by Tenneco, to be merged with J. I. Case, as Case International.

1985 Allis-Chalmers ceases production and sells farm equipment division; they become Deutz-Allis under German owners.

Ford buys New Holland from Sperry Corp.

1987 Caterpillar re-introduces rubber-tracked Challenger Series farm tractors.

3 labour hours required to produce 100 bushels (3 acres) of wheat with tractor, 10.6m (35ft) sweep disk, 9m (30ft) drill, 7.6m (25ft) self-propelled combine and trucks.

1990 AGCO, formed to buy Deutz-Allis Corp. from German owners Klockner-Humboldt-Deutz, begins manufacturing and selling farm equipment under the AGCO Allis and Gleaner brand names.

Ford New Holland sold to Fiat Group; deal states New Holland can use Ford name until 2001.

1991 AGCO purchases the White tractor business from Allied Products.

1992 AGCO purchases the White-New Idea business.

1993 AGCO purchases the North American distribution rights to Massey Ferguson products.

Deere celebrates 75 years in tractor business.

Massey Ferguson's parent company Varity sells North American Massey Ferguson distribution rights to Allis-Gleaner Corporation (AGCO).

1994 AGCO purchases the worldwide holdings of Massey Ferguson.

Tenneco forms Case Corporation, selling Case IH machines.

Varity sells Massey Ferguson division to AGCO, but keeps Perkins engine division.

1996 Case enters the space age with its new Advanced Farming Systems (AFS).

1999 Case and New Holland complete merger to become CNH.

2000 Revised John Deere leaping deer logo is introduced.

2002 AGCO purchases the design, assembly and marketing assets for Caterpillar's tracked Challenger tractors.

AGCO closes the Massey Ferguson factory at Banner Lane in Coventry. The office remains as AGCO's headquarters until 2006.

2004 AGCO aquires Finnish producer Valtra, along with related SISU Diesel engine manufacturer.

2006 The JCB Dieselmax breaks the world diesel-powered land speed record, powered by a pair of specially tuned JCB444 diesel engines.

2009 New Holland unveils the experimental hydrogen-powered NH2 tractor, a 106hp working prototype that operates almost silently with only heat and water emissions.

2011 SAME Deutz-Fahr produces its one millionth tractor.

2012 John Deere celebrates 175 years with a new record of $3.1 billion net profit.

2013 CNH Industrial formed by merger of Fiat Industrial SpA and CNH Global NV.

Index

Index

ACKNOWLEDGEMENTS
The author would like to
thank the following people
and organizations for their
contribution and help.

Andrew Morland Collection:
an archive of pictures we
could never have done
without. Wisconsin Historical
Society, Wisconsin, USA.
North Dakota State
University, Fargo, ND, USA.
Courtesy of Caterpillar: 31;
33TL; 34BL; 39BL. Finning
UK Ltd: 39TR; 40. Courtesy
of John Deere: 92TR, BL;
93TL; BR; 94TR, TC, TL,
BL. The Henry Ford
Collection: 11T. US Library
of Congress, Washington
DC, USA. US National
Archives, Washington DC,
USA. The Ulster Folk &
Transport Museum, Cultra,
Holywood, Co. Down,
Northern Ireland. University
of Guelph Library, Guelph,
Ontario, Canada: 112BL;
113TL, TC, TR; 114TL, TC,
TR; 115TL, CR. G. Bryan
Jones Ltd for supplying
SAME and Lamborghini
tractors. Mirco De Cet
Photographic Collection.
AGCO; CASE IH; Caterpillar
Inc; Ford Photographic
Library, Warley, England;
J C Bamford Excavators Ltd,
Rocester, England; Landini
SpA, Italy; Massey Ferguson,
England; McCormick
Tractors International Ltd.,
Doncaster, England; New
Holland, UK; Renault, UK;
SAME, Italy; Valmet, Finland.